back

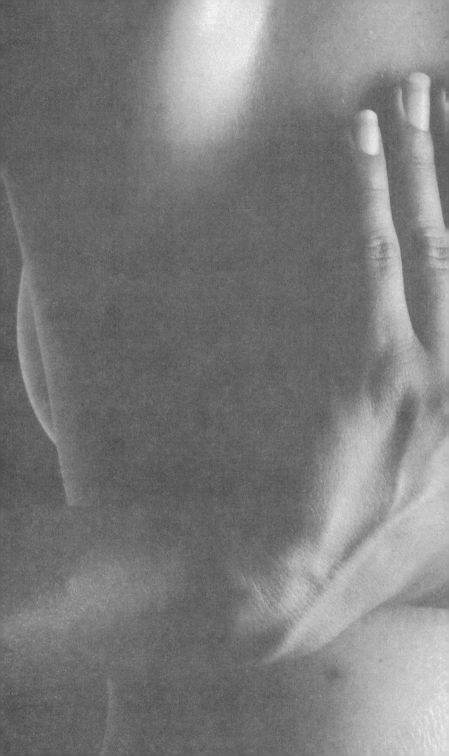

back

Dr. Anthony Campbell

Newleaf

First published by

NEWLEAF

an imprint of

GILL & MACMILLAN LTD

Hume Avenue, Park West, Dublin 12

with associated companies throughout the world

www.gillmacmillan.ie

ISBN 0 7171 3266 8

A CIP catalogue record for this book is available from the British Library.

Note from the publisher
Information given in this book is not intended to be
taken as a replacement for medical advice. Any person with
a condition requiring medical attention should consult a
qualified medical practitioner or therapist.

This book was conceived, designed and produced by

THE IVY PRESS LIMITED

ART DIRECTOR Peter Bridgewater
PUBLISHER Sophie Collins
EDITORIAL DIRECTOR Steve Luck
DESIGNER Jane Lanaway
PROJECT EDITOR Georga Godwin
DTP DESIGNER Chris Lanaway
MEDICAL ILLUSTRATOR Michael Courtney

Printed in Spain by Graficomo, S.A.

Contents

Introduction

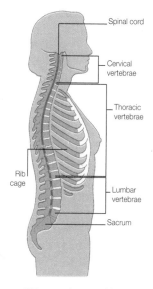

This is a side view of the spine showing anatomical features and normal curvatures.

Spinal cord
Cervical vertebrae
Thoracic vertebrae
Rib cage
Lumbar vertebrae
Sacrum

Back and neck pain is, regrettably, very common. No fewer than 80 per cent of us can expect to suffer at least one episode during our active lives, and back disorders are one of the commonest reasons for absence from work. The annual cost to society in the USA from back pain is estimated to be between $20 and $50 billion. Back symptoms are the likeliest cause of disability in patients under 45 years of age. In one survey, 50 per cent of working adults reported having a back injury each year, and approximately 1 per cent of the US population is chronically disabled because of back pain.

Figures for other industrialized countries are similar. Moreover, the incidence appears to be increasing. In Britain, outpatient attendances for back pain rose fivefold during the decade to 1993, and the number of days of incapacity for which social security benefits were paid more than doubled during the same period. However, it is uncertain whether this is a true increase in the occurrence of back disorders or an increased tendency for people to report their symptoms.

A recent study of two postal surveys in Britain, conducted 10 years apart, suggests that there has been no increase in severe back disease but a marked increase in the apparent prevalence of less disabling back pain. This, in turn, is probably due to cultural changes, which make people more aware of minor back symptoms and more willing to seek help for them; back pain is now more acceptable as a reason for absence from work. The researchers conclude that the solution to the growing economic burden from back pain may lie rather in modifying people's attitudes and behaviour than in reducing physical stresses on the spine.

Back pain can take many forms. A young person may suffer a 'stiff neck' on waking one morning: it comes on without warning, for no apparent reason, and it disappears as mysteriously after a few days. A middle-aged man digs his garden one Sunday in spring, and wakes up next morning to find his lower back stiff and painful, recognizing what has happened because he has suffered from 'lumbago' before. A woman loads shopping into her car, twisting her back awkwardly as she does so, and that evening experiences a dull ache in the lower part of her back, which persists all the following day. On the second morning she wakes to find that her back is so stiff that she can't put on her shoes and she is in severe constant pain. Later the pain eases somewhat in her back but now she begins to feel pain in her leg – sciatica. A fit young man of 30 bends forward to pick something off the ground and suddenly his back 'locks' so that he cannot move or straighten himself.

These are examples of just some of the ways in which pain can occur in your back; there are innumerable possible variations.

In many cases the cause of back pain remains unknown, in spite of the sophisticated methods of examining the spine, such as CT and MRI scans, that are now available. To the average patient this may appear surprising. It may seem that the question 'Why have I got this pain?' ought to be reasonably easy to answer. Having examined the patient, and perhaps taken an X-ray or arranged a scan, the doctor ought to able to say what is wrong, such as a 'slipped disc' or 'trapped nerve', without too much difficulty. After all, the back is a physical structure, capable of being understood in mechanical terms, so why are satisfactory explanations for back pain not always forthcoming?

There are several reasons. First, there is the plain fact that there are still many things we don't know about how the back works. Second, there is the sheer complexity of the back as a structure; it is made up of dozens of bones, ligaments, joints, muscles, nerves and blood vessels. And because of its central position in the body, almost anything that affects any other part of the body may have a secondary, indirect effect on the back as well. It is hardly ever justifiable to isolate one component and then say it is the sole cause of a patient's symptoms. An x-ray, for example, may show that a pair of adjacent vertebrae have ceased to move normally on each other because of disease. But this may not be what is causing the pain; the source of discomfort may be in the vertebral joints further up the spine, which are moving excessively in order to compensate for the lack of mobility below.

It isn't just the mechanical complexity of the spine that makes simplistic assumptions inadequate. We also have to consider the issue of how pain is felt and perceived, and this, too, is more complicated than is sometimes realized. A patient's expectations, social circumstances and mood may all contribute to the amount of pain that is felt.

Medical knowledge about back pain is still developing rapidly. For instance, not until well into the 20th century was it recognized that sciatica (pain in the leg) is often due to pressure on a nerve root in the spine. Another example: until quite recently it was customary to prescribe bed rest for people with recent onset of back pain, on the assumption that this would help the spine to heal. The modern understanding, in contrast, is that patients should keep as mobile as possible. These are just two of the ways in which views about the back have changed in recent years. And they are still changing. Doctors are often provided with what are known as clinical practice guidelines for managing various disorders. These do exist for back pain that has come on recently (acute back pain), but not for long-term (chronic) low back pain, because this is a much more complex problem.

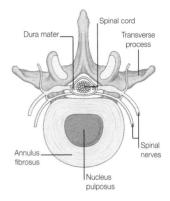

The top view in cross section of a healthy vectrebral spine.

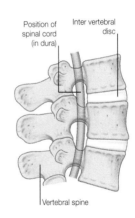

This is an image of the lumbar spine showing the side view.

For all these reasons, diagnosing and treating back pain may be difficult. The best way to start is usually by consulting your family doctor, who is best placed to assess your situation from all aspects. Your doctor may choose to refer you to a specialist, who may be a neurosurgeon, an orthopaedic surgeon, a gynaecologist, a physiotherapist or even a psychotherapist. However, because conventional medicine doesn't always provide a satisfactory solution, more and more people these days are turning to complementary medicine. Doctors themselves are becoming increasingly sympathetic to some of these alternatives; indeed, some now practise them themselves. The boundary between orthodox and complementary treatments is becoming increasingly blurred. Many patients do report that some unorthodox treatments help them, but objective evidence of the effectiveness of these therapies is still scarce. This is yet another area in which more research is needed.

In the meantime, however, many of us need to find ways of coping with the symptoms that we may have and this book aims to answer the common questions that people in this situation often ask.

Prevention

This section is relevant for everyone. If you have never had back pain, it tells you how to reduce your chances of becoming one of the many sufferers; if you have had back pain in the past, it offers suggestions that may help you prevent another attack.

Q

1 What advice can you give to avoid back pain?

Certain basic precautions should be taken by everyone, whether or not they have a history of back pain. These apply particularly to lifting and carrying. Always make sure that you have a firm foothold. Above all, do not stoop, bend your knees and keep your hips lower than your head. This often means squatting, so if your thigh muscles are not sufficiently strong, you should build them up by means of exercises. Get a good grip on what you are lifting and raise it smoothly, without jerking. If you have to turn while holding a weight, do so by swivelling your feet instead of twisting your back. Do not carry even light objects at arms' length in front of you; always bring them close into your body to minimize the leverage that the load exerts on your spine. Finally, when you reach your destination with a load, squat or kneel to place it down; do not stoop.

You should always squat to lift weights, keeping your knees bent instead of stooping.

If you have to move furniture or other heavy objects, give some thought to the most sensible way of doing so first. Plan how you can lift and carry your load safely. Heavy articles should be manoeuvred onto low trolleys if possible before they are moved, but if this cannot be done, at least avoid pushing them from their upper part; usually it is better to sit on the floor and push them with your feet.

You must divide loads into two to spread the strain on your back evenly.

If you are walking back from the shops with a load, divide it equally into two and carry part in each hand rather than all in one hand; alternatively, use a wheeled shopping basket, although many of these are poorly designed and may themselves cause back strain. If you are going to be loading shopping into a car, avoid over-filling the bags; it is better to have many bags with fairly light loads than fewer heavier ones; and carry heavy bags close to your body.

Many older people find difficulty with any kind of work that involves looking upwards for longish periods. Putting up curtains, fixing lights and especially painting ceilings are particularly hazardous. It is not at all uncommon for patients to report pain in their necks, shoulders or arms for several days or even longer after these activities. Moreover, a lot of older people become giddy when looking upwards, so standing on a ladder to work above your head is potentially dangerous.

Q 2 I have had several episodes of backache in the past although I'm all right at present. What can I do to prevent a recurrence?

The main thing to do is to look after your back. This means learning the correct way of lifting, getting in and out of cars, carrying out your daily activities and so forth; many of these points are addressed in other questions in this section. Maintaining a reasonable level of physical fitness and not becoming overweight (or losing weight if you are already overweight) will also help. You may also have considered the idea of carrying out specific exercises designed to strengthen your back and prevent recurrence, (see Q 12).

However, it is important to keep a sense of proportion in all this. You should not become unduly apprehensive about your back or turn yourself into a hypochondriac, constantly fearing that this or that incautious movement will bring on a renewed attack of pain. The fact is that many episodes of back pain come on out of the blue for no apparent reason or following a seemingly trivial event, such as bending down or stepping down suddenly off a kerb. In such cases the pain is probably the final stage in a process of degeneration that has been going on for years, and if this is the case, little or nothing could have been done to prevent it. At the same time, there is no reason to be fatalistic about the situation; much can be done both to minimize the chances of a recurrence and, equally important, to cope with one if it does happen.

If you've had several attacks of back pain in the past you may know what precipitated them, and therefore what to avoid, and you should also know what helped to alleviate them. For example, if you found a TENS machine helpful (see Q 30), make sure you have one handy, with fresh batteries. ●

My local hospital has something called a back school. What is this and what does it do?

Some hospitals have developed what are called 'back schools' to educate people about their backs and to help them prevent recurrence of pain. In a typical back school, a small group of patients is given instruction over several sessions by an experienced physiotherapist.

Participants are taught about the structure and function of the back and are educated in the best way to sit, stand, lift and so on. The idea is to increase the patients' understanding of their particular problems, to allow them to meet and gain support from fellow sufferers, to teach them how to use their backs with greater efficiency and less strains and through exercise to strengthen the abdominal muscles as a protection against further damage to the back.

The effectiveness of back schools in preventing further attacks of back pain, however, is still a matter of controversy; one recent study in the USA, involving some 4000 postal workers, showed no long-term benefit from instruction of this kind; in fact, there were more injuries reported in the group who were taught, perhaps because they were more inclined to report any symptoms they might have. That said, a back school can be an effective way to re-educate you into safe posture.

Q **4** I've been told that a harder bed can prevent backache. Is this true?

It is not necessarily true. It is often said that the harder the bed, the better, and so-called orthopaedic beds are marketed with this in mind, but it is certainly not true for everyone. In fact, a very hard bed allows parts of your spine to sag because they are unsupported. What you require is a bed that supports your spine evenly and allows you to lie in such a position that your spine is as level as possible. A fairly soft bed often does this more effectively than a hard one, but the base of the bed should not be sprung excessively. What you certainly do not want is a sagging mattress that makes your spine curve as it does in a hammock.

Unfortunately, there is really no substitute for trial and error. A few minutes spent lying on a mattress in a showroom are not really adequate – you need to see how you are after a night's sleep. If you wake up feeling worse than when you went to bed, something is probably wrong. You should not persevere with a mattress that is uncomfortable.

Reading in bed is difficult to do without straining your neck and back. If you lie on your back with your head propped up on pillows, you are liable to strain your neck; if you sit upright with your legs in front of you, there will probably be strain on your lumbar spine. It may be better to avoid reading in bed altogether, or at least to change position often.

As a general rule, two pillows are best, but a lot depends on the thickness and resilience of the pillows themselves. Foam-filled pillows are undesirable because they do not give. Although down pillows have a better consistency, some people are allergic to them. Whether you use one pillow or two, the critical consideration is that your head, neck and spine should be in approximately a straight line when you are lying on your side. You do not want your head either to droop appreciably or to be raised excessively – both these positions place an undue strain on your neck. To achieve this, you should make sure that the pillow is tucked well into the space between your head and your neck, with no unsupported gap between the two. Special horseshoe-shaped pillows are available to provide this kind of support, but you can achieve the same effect with a rolled-up towel.

Any pillows you use at night should be used to keep your spine as level as possible. Use your pillow to support your neck rather than your head.

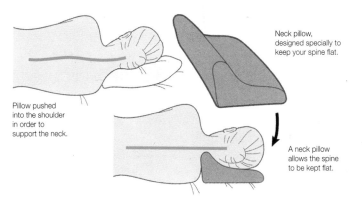

Neck pillow, designed specially to keep your spine flat.

Pillow pushed into the shoulder in order to support the neck.

A neck pillow allows the spine to be kept flat.

Q

6 I'm aged 50 and I have had some back trouble in the past. I'm afraid that having intercourse will damage my back. What's the best way of avoiding this?

Sexual intercourse is still possible if you have spine or back problems.

It is true that intercourse can cause back strain. The solution is to experiment with different positions to find those that are comfortable for you and your partner, perhaps with the help of a strategically placed pillow. A reasonably firm bed will help, although there is no need for it to be excessively hard, and the partner prone to back trouble should take up an S-shaped posture, the other partner fitting into this as required. If any pain is experienced during intercourse, this should be signalled at once. With a little experimentation it should be possible to find a position that is satisfactory to both. If this is not the case, you could ask your doctor to refer you for advice from an occupational therapist (*see Q 51*) or physiotherapist.

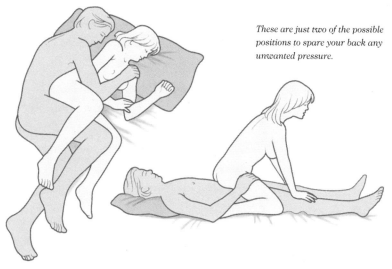

These are just two of the possible positions to spare your back any unwanted pressure.

7 My husband is a keen gardener but he often says that his backache is worse after he has been working in the garden. What can he do to prevent this?

Gardening often causes problems, especially in spring when people resume work after the winter. Digging involves a lot of strain on the back and so does mowing with a non-motorized mower. It is important to stop frequently for a rest and change of posture – do not go on too long with one specific task and stop before you feel tired. If you have a large lawn, for example, break off before you've finished, do something else for a time and then go back to the mowing.

Weeding, although not heavy work, often causes people to stoop. Instead, try kneeling on a rubber mat.

The standard wheelbarrow design is ergonomically unsound and places a lot of strain on the lower back. To avoid this, do not overload the wheelbarrow. It is better to make a few extra trips than to risk serious back injury.

In winter, be particularly careful when you are clearing snow, not only because of the lifting that is involved but because of the danger of slipping on ice. Clearing snow can be dangerous for the heart as well as for the back, because the type of exercise that is involved (known as isometric) can affect blood flow in the heart muscle.

You should not bend over when gardening, either crouch with your knees bent or use special kneeling aparatus.

Q

8 What can I do to prevent back and neck problems from spending long hours at my desk?

You may have less than ideal control over the conditions under which you work, but it is important to do what you can to make sure they are as favourable as possible.

If you spend most of your time sitting at a desk, perhaps working at a keyboard, it is important to keep your posture in mind. As in all kinds of sitting, slouching is undesirable. You should aim for an upright but not super-erect posture. To achieve this, the height of your desk in relation to your chair is critical. Much pain in the neck and shoulders arises because the desk is too high or too low. Your elbows should be at 90 degrees with your forearms parallel to the floor when sitting at your desk. You should also ensure that you are sitting neither too near nor too far from your work surface. If you have difficulty in achieving the right position you should get a good chair that allows for adequate adjustment. Some people find a 'back chair' is excellent. This is a radically different type of chair, in which you half-sit, half-kneel: the seat is angled slightly forwards and part of your body weight is taken by a

A back chair gives a better posture at work and can stop you having problems in the future. They also relieve any current back pain.

second support below the knees. These chairs tend to be expensive but anyone whose work or leisure involves a lot of sitting should consider them.

Even the best posture, however, is going to become uncomfortable if maintained for too long. Every 20 to 30 minutes, therefore, you should move about, look away from your work, perhaps get up for a moment. Shrugging the shoulders often gives relief from neck and shoulder stiffness. Also stretch your fingers and perform some simple neck exercises (*see Q 13*). It is important that you do these things by the clock, rather than waiting for discomfort.

If you have to do a lot of copying, do not lay the material you have to copy on the desk. Instead, prop it up in front of you so that you do not have to keep looking down.

Do not use the telephone with it tucked between your neck and shoulder and your head drastically bent to one side as this puts unacceptable pressure on your neck. Stop what you are doing in order to leave your hands free, or else get a hands-free phone.

Even office work can involve a surprising amount of quite heavy lifting. Papers and files can be very heavy and add to the daily toll on your back.

Always remember the golden rule about not applying strain to a bent back, and especially a bent twisted back. Squat before lifting and get the load as close to your body as possible, in other words do not carry any heavy loads with outstretched arms, and distribute heavy loads evenly.

Q

I'm due to fly to Australia shortly and I have had problems with my back when travelling in the past. What can I do to make sure I arrive at my destination free of back pain?

Travelling by air is liable to impose strains on your back. Even before you arrive on the aircraft you have to manage to transport your luggage to the check-in desk. It is best to have several small suitcases rather than one or two large ones and to carry some of the load in each hand rather than all on one side. Once you have checked in, you should avoid burdening yourself with heavy items, such as bottles of duty-free drink. If you must carry heavy items as hand luggage, at least provide yourself with a portable, wheeled luggage carrier. Once on the plane, ask for help with lifting things into an overhead locker.

If you suffer with severe pain or find walking difficult, notify the airline in advance and ask for a wheelchair or buggy to be made available at both departure and destination. Airlines are usually helpful about this (and you may get through the formalities faster than the more mobile passengers).

Once on board the plane, you are likely to find limited room for your legs. Again, if you notify the airline in advance they may be able to accommodate you in a seat that allows you to move more freely. On long flights get out of your seat and move about from time to time – at least every hour. Remember to check your seat position and, if necessary, ask the attendant for a cushion to place behind your back.

The vast majority of back sufferers – and, indeed, many people who do not normally have back trouble – report that driving a car causes them discomfort. There are several reasons for this.

The most obvious source of difficulty is the seat. Few cars have seats that are designed to give the right support for the back, and even well-designed car seats cannot suit everybody. However, the most common fault is lack of lumbar (lower back) support, followed by too short a seat giving inadequate support for the thighs. Unfortunately, a few minutes spent sitting in a car in a showroom are not usually enough to tell you whether it will be comfortable in the long term. It is sometimes possible to improve a less than ideal seat with a cushion or a back support.

Your position in relation to the controls is also important. The height and angle of the steering wheel and its distance from your body can all affect you, as can the height and position of the pedals. It may be possible to alter the angle of the steering wheel and the height of the seat, if not you may want to consider using a cushion to attain the right height.

A heavy clutch is a source of back pain for some people, especially with town driving. In such cases it may well help to

Adaquate thigh, back and head support is needed when driving. Check the way you sit to drive against this illustration.

change to a car with automatic transmission. A less obvious source of trouble is low-frequency vibration from the engine. This is a somewhat contentious subject, but there have been some reports that noise of this kind can aggravate patients' symptoms.

Many people experience pain and stiffness in the neck, shoulders and arms after driving for a time. Apart from making sure that your posture is as good as it can be, whenever the traffic comes to a halt, turn your head, relax your grip on the steering wheel and shrug your shoulders. On the motorway, stop frequently at service stations – at least every hour – and get out of the car to stretch your muscles.

Factors such as faulty posture and badly designed seating also interact with the psychological tension that so many of us suffer from when driving. This causes us to grip the steering wheel more tightly and tighten the muscles of our necks and shoulders. It helps to make a conscious decision to improve our driving skills and to cultivate a detached attitude to incidents that occur while we are at the wheel.

As with other load-carrying activities, it is important to take care when putting luggage into the car or taking it out. Avoid large, heavy suitcases and try to choose a car that does not have a high sill over which things have to be lifted. Avoid turning around in the car to lift things out of the back seat and try to get out of the car without twisting your spine, training yourself to swivel your whole body around on your buttocks before standing up.

I get quite a lot of backache and I'm pretty unfit. I have been told that getting fit would help my back. Is it true and how should I go about it?

Regular exercise is important, not only for the heart and circulation, but also for the back. It improves the strength of bones and so helps prevent osteoporosis. Exercise is also one of the best methods of counteracting tension and having good tone in the abdominal muscles is beneficial for the spine.

These benefits apply regardless of age; however, it is only sensible to take your age, health and present state of fitness, or unfitness, into account. If you are over 40, suffer from high blood pressure, diabetes or any long-term disease of heart or lungs, or are seriously overweight, you should consult your family doctor before starting an exercise programme. The main exception to this is walking, which is suitable for almost everyone except those with very serious illnesses. Even with this, however, you should start gently (just a few minutes a day) and build up slowly.

The best forms of exercise are those that involve large muscle groups, that is, the muscles of the back and legs. These include walking, jogging and running, swimming and cycling. Tennis and golf are also good, but may cause problems to back sufferers because

It is important to remember when swimming that breast stroke can strain your back. Keep your head down and your body as flat as possible to make sure you do not have any problems later.

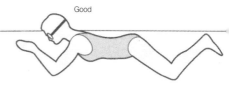

Bad   Good

they tend to involve twisting movements. Dancing is good exercise, although some people do experience backache as a consequence.

Most of the health benefits can be gained by performing exercise of moderate intensity, such as walking briskly at three to four miles an hour on most days of the week. It is important to warm up and slow down gradually at the end of the exercise period. You should also perform stretching movements before and after exercise, but do not force yourself to touch your toes. Bending forward in this way may squeeze disc material out of your spine.

Swimming is often recommended for back sufferers because water supporting the body and limbs reduces strain on the joints and ligaments. However, not all the different strokes are equally good. Breast-stroke, if done well, is fine, but keeping your head out of the water all the time places a strain on the neck and back muscles. If you do the crawl, try to breathe to each side alternately rather than to only one side. Safest is backstroke.

Walking is the most trouble-free form of exercise, but if you need to carry a load, use a belt so that the weight is carried by your pelvis. Make sure that dogs are trained to walk on a lead without pulling. Jogging and running can help you to get fit quickly, but make sure that your footwear is suitable, especially if you exercise on pavements. Also, pay attention to your running posture; many people carry their arms too high, imposing a strain on their upper back and neck.

12 Are there any
specific exercises
that will help to
prevent recurring
back pain?

Almost every popular book on back pain places a lot of emphasis on the value of specific back exercises. Unfortunately, there is not a great deal of hard evidence to show that such exercises are truly beneficial and still less about which kinds of exercise are best. If you want to try an exercise programme you should certainly do so, but if you find the idea unappealing you shouldn't feel that you are necessarily slowing down your recovery or making a relapse more likely if you do not do exercises, unless they are suggested by your doctor or therapist.

You should also keep in mind that exercises can do harm as well as good. A moderate amount of discomfort during the actual performance of the exercise is not necessarily a reason to stop, but exercising to the point of pain is definitely undesirable. Certain exercises in particular should be avoided. Toe touching with straight legs, and especially bouncing in an attempt to overcome stiffness in the back, puts an excessive strain on the lumbar region of the spine. Sit-ups with straight legs and lifting straight legs off the floor from the hips should likewise be avoided, since they hollow the lumbar spine putting additional strain on the region. Neck mobility is certainly worth preserving, but swinging the head round in circles is not a good thing to do; move gently and carefully.

13 Can you suggest some particular exercises that would be good for the neck?

This exercise should be done first. To exercise the upper joints of your neck, sit in a chair with your hands in your lap or on the sides of the chair; then glide your head and neck from side to side 4 or 5 times like a Balinese dancer. (Look in the mirror or get a physiotherapist to demonstrate this, if in doubt.) As a general neck exercise look up to the ceiling and down to the floor. Repeat 4 or 5 times .

With neck bent forwards, look to the left and right. Repeat 4 or 5 times in each direction. Some patients with one-sided or shoulder pain experience relief if the head is bent towards the affected side a few times. But if this makes the pain worse, or pain moves farther down the arm, do not persevere.

To strengthen the muscles on the front of your neck, lie on your back on the floor with knees bent and your head on two pillows. Then raise your head smoothly, bringing your chin down towards your chest as far as possible, before lowering your head gently back onto the pillow. Repeat 10–20 times. Alternatively, sit in a chair and place your hands, with fingers interlocked, against your forehead. Slowly increase the pressure of your head against your hands, countering it with increased pressure from your hands; hold for 2–3 minutes.

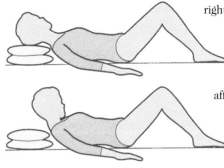

The correct way to exercise your neck and abdominal muscles is to do so carefully. Keep the small of your back on the ground and move slowly raising your shoulders rather than your head.

Q

14 Are there any particular exercises that would be good for the thoracic spine (chest region)?

The aim of exercises for this region is to try to maintain or improve mobility, especially in the upper part of the spine where many middle-aged and elderly people become stiff and immobile.

Stand in a doorway with your arms raised above your head. Place your palms on each side of the doorway so that your body forms a 'Y' shape. Keeping your elbows straight, press your chest (not the stomach region) through the doorway. This helps to relieve pain in the lower part of the neck at the back and in the top of the shoulders (the yoke area).

This exercise can be made more vigorous if the hands are turned through 90 degrees so that the palms face downwards during the bracing back of the shoulders. Gradually increase the length of time you hold the posture, but 5 to 10 repetitions are enough.

For pain in the yoke area, try lying on your back on the floor with a rolled-up towel under and running along the upper third of the thoracic spine. Allow your head and the base of your neck to fall towards the floor and rest there for a few minutes. Also, while sitting, shrug your shoulders up towards your ears and hold them there for a few seconds.

To maintain movement in the thoracic spine, kneel on the floor with your buttocks on your heels, rest your forehead on the floor and place your hands, palm upwards, on the floor beside your feet. Raise your head and trunk to a horizontal position, bracing your shoulders back as you do so. Breathe normally and do not hollow your lower back.

15 Are there any particular exercises that would be good for the lumbar spine?

Throughout the day, whether sitting or standing, make a conscious effort to remember to check your posture. Try either increasing or decreasing the amount of hollow in your lumbar region (lordosis) until you feel comfortable. Remember that there is no one 'correct' posture, experiment and see what is most comfortable for you.

Lying on your back with knees bent, flatten your back so that your lumbar spine touches the floor and your pelvis comes forward (a) and at the same time, tighten your buttock muscles (b). This exercise can be done standing as well as lying and should be performed at intervals throughout the day.

Illustrations (a) and (b) show an exercise to tone the buttock muscles and flatten the lumbar spine.

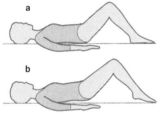

Lie on the floor and stretch downwards with your left leg as far as you can, tilting your pelvis at the same time. Hold for a few seconds and then relax. Repeat the same exercise with the right leg. Repeat 5 to 10 times each side.

Lying on the floor, bend your left leg and hip and pull your knee gently towards your chest with your hands. Hold for a few seconds and relax. Repeat with the right leg. This exercise stretches the psoas muscles – big muscles on the inner side of the spine that move the thighs. Repeat 5 to 10 times each side.

Images (c) and (d) show an exercise to strengthen the abdominal muscles.

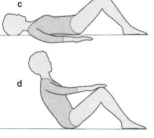

Lie on your back with your hips and knees bent and your arms relaxed by your side (c). Smoothly, and without holding your breath or jerking your

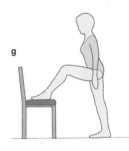

These exercises stretch the hamstrings and strengthen the back muscles. Remeber to do them slowly and carefully with no jerking.

head and neck, raise your head and the upper part of your body off the floor and hold the posture for a few seconds (d), before lowering them gently back to the floor. This exercise is intended to tone up your abdominal muscles. You can also practise turning your trunk as you lift it so as to bring each shoulder in turn towards your knees; this action will exercise slightly different parts of the muscles. Repeat 5 to 10 times. Stop immediately if this causes pain.

Lie face downwards on a strong table; the edge of the table should be underneath your hips, and you should use a folded towel for padding (i). Grip the top or sides of the table with your hands and then lift your legs and feet off the floor and raise them as high in the air as you can (j). Lower your feet smoothly to the ground. This exercise strengthens the big muscles of the back. Repeat 5 to 10 times.

Place your left heel on a chair or stool and, keeping your left leg straight, slide your hands down your leg as far as you can, but do not bounce or strain to reach further (e). Repeat with your right leg. This exercise stretches the hamstring muscles: the big muscles located on the back of the thigh. Repeat 5 to 10 times.

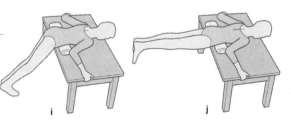

Place one foot on a chair (f) with the other leg as far behind you as is comfortable (g). Slowly bend the leg that is resting on the chair (e), then straighten that leg and slowly bend the leg on which your are standing (h). Swap legs and repeat. (This is another exercise for the psoas muscles.) Repeat 5 to 10 times with each leg. ●

Q

16 Will practising yoga help to prevent back problems?

Yoga is an ancient Indian system for conditioning the body and mind. In its full, original form it consists of a number of techniques for mental and spiritual as well as physical development, but most Westerners approach it mainly on the physical level. However, it is a mistake to think of it primarily as a form of exercise and it will not do anything for your physical fitness. It can, however, improve flexibility and relaxation.

The physical techniques of yoga are known as asanas, which means postures rather than exercises. Each posture is supposed to be held for a certain time. Some are very simple to achieve (one consists merely of lying flat on one's back), but others, such as head stands, are extremely demanding and require years of practice to perform properly and safely. Most Indian yoga practitioners began their study as children and are therefore very supple, but it is quite different for a middle-aged Westerner who has no previous experience of yoga.

Provided the teacher is experienced and sensible, yoga should do no harm to back sufferers and some undoubtedly find it beneficial. But it is very important not to strain to achieve difficult positions, and some of the classic postures should not be attempted at all by the middle-aged or elderly, or by anyone with a history of spinal trouble. This applies particularly to the shoulder stand, which forces the neck into a potentially dangerous position. Movements involving twisting the spine should also be attempted cautiously or not at all.

There are many books on yoga available, but be wary of attempting more complex positions if you have back trouble. It is always advisable to practise them first under expert supervision.

17 I used to cycle a lot when I was younger but I haven't done so for years. I'm thinking of starting again but would it be harmful for my back?

Cycling can be a surprisingly good form of exercise for back sufferers. Some people find that they can cycle even when walking is painful. It has two advantages over many other kinds of exercise: your weight is supported, and because you tend to lean forward, the lines of stress in your spine may be different from what is usual when you walk. Patients who suffer from narrowing of the spinal canal in the lumbar spine, causing neurogenic claudication (*see Q 46*), sometimes find that they can cycle quite comfortably, despite walking being highly painful for them.

It is important to choose a suitable bike. Many people with back pain tend to think that an upright posture, as provided by the old-fashioned 'sit up and beg' type of bicycle, will automatically be best for them, but this has the disadvantage that road shocks are transmitted up the nearly vertical spine. A more sloping posture is therefore likely to be better, but drop handlebars are not recommended, because they require you to raise your head, which puts a strain on the neck. Probably the best choice is one of the 'all-terrain' machines, which will have fatter tyres and perhaps a sprung saddle. You can also mount the saddle on a telescopic seat pin for better shock absorption. Try out a few different types to see what is right for you.

Whatever kind of bike you buy, make sure it is the right size. If the frame is too small, you will be hunched up, but if it is too large, you will have to reach too far forwards to grasp the handlebars. It is difficult to judge the right size if you do not know much about bikes, so it is worth paying a bit extra to buy from a knowledgeable dealer who can give you accurate advice.

Even if the bicycle is the right size for you, you must make sure that it is correctly adjusted. Handlebars set too low cause neck pain. If the saddle is too high you will rock your pelvis from side to side as you pedal and this can cause backache. The saddle angle is also critical – usually, dead level is best, but it pays to experiment.

When things go wrong

Back problems are very common; at least half of us can expect to experience back pain of one kind or another during our lives. This section of the book provides information about coping with back pain, diagnosis and the different kinds of treatment that are available.

First Attack

The first attack of back or neck pain can be a frightening experience. The pain may be severe and you may be unable to move or straighten up. This section looks at the possible causes of back pain, and explains what happens to your back during certain illnesses and what you can do to help relieve the symptoms, especially the pain.

18 I'm just getting over my first attack of back pain, which lasted two weeks. Does this mean I had a slipped disc?

You did not necessarily slip a disc – there are many other possible causes of back pain. The term 'slipped disc' is pretty misleading in any case, because discs (the shock absorbers situated between the vertebrae) cannot really slip.

The vertebrae mostly are formed of a cylindrical block of bone, known as the body of the vertebra, backed by a triangle of bone which forms an arch housing the spinal cord or, in the lumbar region, the nerves coming off the cord. The sequence of these arches along the spine constitutes a tunnel for the spinal cord (the spinal canal). The discs are found between the cylinders. They consist of a ring of tough material called the annulus fibrosus, which is bonded firmly to the vertebra above and the vertebra below. In the middle of the ring, there is a jelly-like substance called the nucleus pulposus. As we age, the material of the ring of the disc may become worn and then a sudden movement, only a cough or a stumble, causes the jelly-like substance to squeeze out through the crack. It may press on a nerve root or the spinal cord, depending on the level at which it occurs. This is called disc prolapse. Usually the disc material goes to either the left or the right, in which case the pain will be felt on that side. Occasionally the leak is massive and central, giving rise to pain on both sides as well as causing other, more serious effects, such as loss of bladder function, but this is rare. Disc material can also escape upwards or downwards; there is then no nerve root pain, only backache. There is no way to put this material back once it has come out. Fortunately, it tends to shrink with time, so most people recover within a few weeks. It is also possible for disc material to bulge, without actually coming out of the disc. Bulges can go back, although it is always possible that they will recur. Often they cause no symptoms at all.

There are three main regions of the spine: the cervical or neck region; the thoracic or chest region; and the lumbar region. Disc prolapse can occur between vertebrae at any level in the spine, but most often it happens in the lumbar region, partly because this part of the spine carries most of the body's weight. The next most common region is the neck, where it may press on the spinal cord. This can be serious because it may lead to weakness in the leg or to other effects in the lower part of the body.

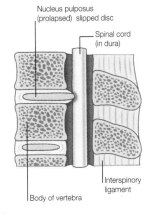

Nucleus pulposus (prolapsed) slipped disc

Spinal cord (in dura)

Interspinory ligament

Body of vertebra

A prolapsed (collapsed) disc is squeezed so that it presses on the spinal cord thus causing pain.

Recently I had an attack of low back pain; after a few days the pain in my back got better but then I had a lot of pain in my leg, which is now going away gradually. Was this sciatica?

Sciatic nerve   Spinal cord

The sciatic nerve runs from the small of the back, lumbar region and down the legs. It is the longest nerve in the body.

What you are describing is a typical attack of sciatica. The term 'sciatica' really means pain in the area supplied by the sciatic nerve, which is mainly the back of the leg, but it is often also used to mean pain in the leg due to pressure on other nerves, such as the femoral nerve, which supplies the front of the thigh. This is often due to disc prolapse (slipped disc), but even though you may feel pain radiating down your leg this does not prove conclusively that there is disc damage; the same symptom can be produced in other ways. However, if there is definite weakness of muscles in the leg, the chances of a disc prolapse are considerably greater.

The way your symptoms developed is quite a common one. At first there is a lot of pain in your back, often with associated spasm of the back muscles. This is a protective reflex and may be so strong that you find it impossible to put on your socks and shoes. After a few days the muscle spasm passes and the pain in the back gets better, but now there is pain in the buttock and leg, which may spread as far down as the foot. The leg may feel cold and there are often pins and needles or numbness. The leg may feel weak and, in the most serious cases, at a later stage the muscles may begin to waste away. Occasionally sciatica occurs without backache, although this is unusual.

I have had one attack of low back pain, which fortunately I have recovered from completely after two weeks. My doctor does not think I need an x-ray, but I want to know what caused the problem. Why does he not want me to have the test?

Your doctor is right. These days it is considered good practice to reduce exposure to radiation as much as possible, so x-rays are done only when absolutely necessary. There are seldom any reasons to x-ray the spine during the first month of pain. The question the doctor has to ask herself is whether an x-ray would alter the treatment or provide any useful information. In your case it is very unlikely that an x-ray would show the cause of your backache. It might show a completely normal spine, it might show 'wear and tear' changes, or it might show loss of disc space, but there would be no way of telling whether these changes had any connection with your recent back pain episode, nor would they say anything useful about the chances of recurrence. Plain x-rays do not show the soft tissues (muscles, ligaments, discs and nerve roots) which are often the cause of temporary back problems. More sophisticated tests, such as MRI scans, are used to see such details, but these are normally done only when surgery is being contemplated.

It is sometimes advisable to x-ray the back during the first attack of backache, but this is usually because the doctor suspects there could be a more serious cause for the pain, such as infection or cancer, rather than the common 'mechanical' cause.

Q

21 My brother has had one attack of back pain, which he recovered from completely. How likely is it that he will have another?

It is extremely difficult to predict a person's chances of having a further attack of back pain; much depends on your age and other environmental factors, such as your work, your sporting activities if any and perhaps your family history. The answer will obviously depend in part on what caused your first attack, but the back is a very complex structure and there are many ways in which pain can arise. The cause of back pain remains unknown in about 85 per cent of cases.

Given that back problems are very common in the population, there is a fair chance, statistically, that it will happen again. On the other hand, one attack does not necessarily make it more likely that you will have a second attack. This may well have been a one-off event, due, for example, to a lot of heavy lifting or some other circumstance that might have caused a disc to prolapse. If you know what precipitated this episode it would be common sense to avoid repeating whatever caused it; otherwise, you should simply follow the general recommendations for looking after your back as outlined in the 'Prevention' section. ●

My son, aged 14, has woken up twice with a stiff neck and he could hardly turn his head. He got over it each time in a couple of days. My doctor isn't worried. What causes these painful attacks in a young person?

This is quite a common thing to happen to young people – it is called acute wry neck. Recovery usually occurs within a few days, with or without treatment, although occasionally it may take 10 days or more. The cause is unknown. Some people move about surprisingly vigorously when asleep and it is possible that they contort their necks into an unnatural position that strains the muscles or joints. The opposite can also occur: during dreaming sleep the muscles are profoundly relaxed, so that the neck can fall into an unnatural position, and this, too, can strain the joints. Either of these mechanisms could slightly displace one of the facet joints in the neck or cause a disc to bulge, owing to a slow shift of disc substance during sleep. There is unlikely to be any pain in the arms or hands.

The duration of pain may be reduced by physical treatment (manipulation, physiotherapy, acupuncture). Local warmth may speed recovery and simple pain-relievers such as paracetamol may help.

Older patients can also suffer from acute wry neck, but there is often underlying osteoarthritis of the facet joints. Gymnasts and people who go in for contact sports may injure their necks by tearing ligaments, joint capsules or small muscles. The duration of pain tends to be longer and there may be pain or numbness in the arms and hands.

23 I had an episode of low back pain last month. It was not too severe but I thought I should go to bed to rest. However, my sister said she thought I should stay up and move about. Who was right?

There is a lot of evidence nowadays to suggest that it is better to keep mobile if you have back pain, so your sister may well be right. By moving about, you keep a lot of normal information coming into your spinal cord, which tends to prevent the central nervous system from focusing too much on pain. If you lie in bed, on the other hand, not much is going on in your muscles and joints and so the pain information has the field to itself, as it were.

Obviously a lot depends on how severe the pain is. If it is very bad, you will be incapable of moving, so going to bed is the only choice. However, bed rest does not actually help, and most patients recover more quickly if they remain up and move about. Obviously you should avoid movements that might have caused the problem in the first place.

Some specialists, particularly in Scandinavia, have gone even further. They now treat acute back pain by means of vigorous back exercises, done repeatedly in sessions lasting an hour. They claim that this can cure many patients' back pain almost immediately, although it is fair to say that this form of treatment has not yet been fully evaluated.

One study in Finland found that neither bed rest nor designed exercises were as beneficial as advising patients to carry on with their normal activities within the limits prescribed by their pain.

My mother, aged 65, was getting a lot of pain in her shoulders and her doctor diagnosed polymyalgia rheumatica and prescribed prednisolone, which I understand is a steroid. I'm worried about this; what is polymyalgia rheumatica and why does it need strong drugs for treatment?

Polymyalgia rheumatica is an important, though fortunately not very common, cause of pain and stiffness in the neck, shoulders and back. It is important partly because it is very easily treated, once diagnosed, but also because it may be associated with another closely related disease, called temporal arteritis, which may lead to blindness if untreated. Polymyalgia rheumatica affects women more than men and usually occurs after the age of 60. It can come on very suddenly and is characterized by severe muscle pain, which may be particularly troublesome in the morning so that patients are barely able to get out of bed. Patients feel unwell, but such complaints may be dismissed as simply part of the ageing process affecting the joints. The disease is diagnosed by a simple blood test.

Corticosteroids relieve the symptoms very rapidly and keep it under control until it clears up spontaneously – usually after some months or years. They act by reducing the activity of the immune system, which is overactive in polymyalgia. The risk of side-effects is outweighed by those of the disease and by the need to relieve the pain it causes. Although it may be necessary to use quite a large dose to start with, this can be reduced as soon as the symptoms come under control and your mother will then remain on a lower maintenance dose.

25 My partner is suffering her first attack of back pain. She is coping pretty well with painkillers and we do not like to trouble our doctor unnecessarily. Is it essential to see him?

It is not always necessary to see your doctor immediately for back pain. It is quite reasonable to wait a few days, taking pain-relievers and see if you get better. There are, however, a few pointers that you need to be aware of.

The first is age: if you are aged under 18 or over 50 and this is your first attack of pain, you probably should see your doctor unless you recover quickly. This is to exclude the rare but serious causes of back pain, such as cancer, that affect people in these age groups particularly. If you have suffered for a long time from back pain, however, the fact that you have passed the age of 50 is not particularly significant.

Next, pain that is much worse when you lie down, or which occurs only at night, is another reason to consult your doctor. These patterns are rather unusual and may suggest a potentially serious cause.

If you have fever, chills or weight loss you should see your doctor, since these are not expected in mechanical back pain; also if there is any disturbance of bladder function or loss of bowel control (short-term constipation is not a cause for alarm).

A recent history of infection, such as a boil or a kidney infection, should prompt you to visit your doctor, as infection sometimes settles in the vertebrae; as should a weak limb. If you are diabetic this, too, is a reason for a consultation.

SUDDEN ONSET OF BACK PAIN

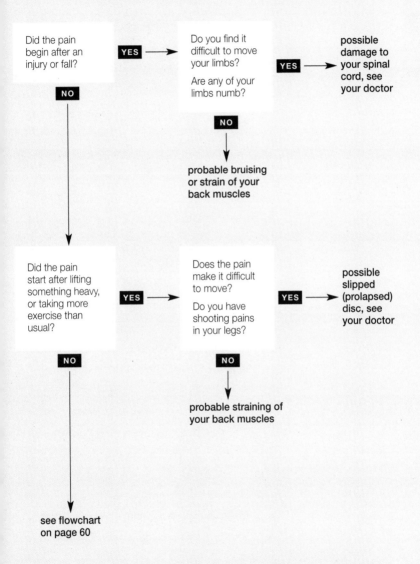

Did the pain begin after an injury or fall?

YES → Do you find it difficult to move your limbs?

Are any of your limbs numb?

YES → possible damage to your spinal cord, see your doctor

NO ↓

probable bruising or strain of your back muscles

NO ↓

Did the pain start after lifting something heavy, or taking more exercise than usual?

YES → Does the pain make it difficult to move?

Do you have shooting pains in your legs?

YES → possible slipped (prolapsed) disc, see your doctor

NO ↓

probable straining of your back muscles

NO ↓

see flowchart on page 60

You should always consult your doctor when suffering from unexplained back pain.

Q 26

I'm in bed with an acute attack of back pain. Which is the best position to lie in?

You should lie in the position in which you feel most comfortable. There is no 'right way' to lie, although lying flat on your back or on your front is unlikely to be good because these positions place a strain on the back. Lying on one side is often best, with a pillow between your knees to prevent your upper thigh from pulling you over and twisting your spine. Whatever position you adopt, the ideal is to keep your whole spine as level as possible, and cushions and pillows should be used to achieve this.

Turning over in bed may be difficult. It is often easiest to draw your knees up and then to allow the weight of your legs to pull you over without twisting your spine. It is easier to do this if you do not have heavy bedclothes on your legs. It is better to keep the room warm and use only a light cover if possible.

Helpful friends and relatives are likely to suggest that your bed is not hard enough and may want to place boards under your mattress to make it rigid. As a rule, this is not necessary: a very hard bed may indeed be more uncomfortable than a softer one. Provided your bed does not actually sag, it should be satisfactory as it is. If it does sag a lot and if you do not have any

An alternative position if you have pain in the front of your thighs. In this position you must make sure that your back is flat to the floor and that any pillow you use supports your neck rather than your head.

boards handy, you could ask someone to pull your mattress onto the floor. For more information on beds, (see Q 4).

Lying flat may not be the most comfortable position. Another possibility, if you are on the floor, is to put a chair at your feet with a pillow on the seat; you can then lie on your back and rest your legs on the chair, so that your hips and knees are flexed at about 90 degrees. This is particularly useful if, as may happen, you have a lot of leg pain in the front of your thigh rather than the back, because in this case the pain will be less if your hip is flexed.

Getting to the lavatory may be a problem. Provided a lavatory is within reasonable distance of your bed and on the same floor, it is best to use it rather than to attempt alternative arrangements. Bed pans and so forth, even if they are available, have more disadvantages than benefits for most people. It is likely that you will be constipated for the first few days. In part this is simply due to inactivity and to the pain itself, which tends to inhibit the normal movements of the bowel. Sometimes, if the pain is severe, your doctor may prescribe drugs from the opium family. These have the side-effect of causing constipation. In any case, it is best not to take any medicine for the constipation but to wait until it sorts itself out naturally, which it will. However, it is sensible to eat plenty of fruit and vegetables and to drink adequate amounts of fluids, both of which will help to counteract any tendency to constipation.

Q

27 I had a lot of pain in the middle of my back for about a week. Now I have developed a rash around my right side. Is this shingles? And, if it is, will the pain go away as the rash clears up? And what about treatment; can my doctor help?

This does sound like shingles, although you should see your doctor to make sure and to obtain some treatment.

Shingles, or herpes zoster, is due to reactivation of the chicken pox virus. At some time in the past, probably when you were a child, you will have had chicken pox. The virus did not entirely disappear from your body, but remained hidden in the nerve cells in the sensory nerve roots of your spine. It may stay there for many years, until, for one reason or another, it becomes active again. This can happen because your immune system is working less well, perhaps because of another infection, perhaps because you are run down, perhaps just because you are getting older. It does not, however, follow exposure to a child who has chicken pox – it is not due to a new infection with the virus but rather to reactivation of a previous one. It can occur at any age but it increases in frequency as people become older.

When the virus is reactivated in this way it does not cause a fresh attack of chicken pox but it does give rise to pain, and it also causes a rash to appear in the skin supplied by the nerve root in question. Sometimes more than one nerve root is affected, giving rise to a larger area of rash. The rash takes the form of blisters at first; these soon break

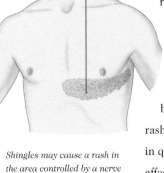

Shingles rash

Shingles may cause a rash in the area controlled by a nerve root. In this illustration it appears on the left of the chest.

down and become pustules, which heal but may leave an area of scarring. Any level of the spine may be affected, though the chest is one of the most common sites. If the eye is involved it is potentially very serious and expert help is needed to prevent damage to the sight, including blindness.

The amount of pain and the severity of the rash, are both variable. Occasionally the pain occurs alone, without the rash, although this is unusual. The pain may precede the rash by a week or 10 days, as in your case, and diagnosis can be difficult at this time. Treatment is directed mainly at controlling the pain. Lotions such as calamine make the rash dry up more quickly.

The main concern in shingles is the development of what is known as post-herpetic neuralgia. In some cases the pain does not clear up when the rash fades but continues for months or even years afterwards. Modern surveys suggest that this is not as frequent as we used to think but it is a difficult problem to alleviate when it does occur; TENS (*see Q 30*) gives relief to some patients with this condition. Various treatments, including antiviral drugs, have been tried in the hope of preventing this complication. If they are to work they must be given as early as possible in the course of the disease, but their effectiveness still remains uncertain.

The good news is that, if you have had one attack of shingles and completely recovered from it, you are highly unlikely to have another.

Q

28 **I'm having my first attack of severe backache. How long is it likely to last?**

By definition, you are suffering from acute back pain, which is defined as back pain lasting for less than three months. The good news is that full recovery occurs in 85 per cent of cases. In most of these the duration is less than six weeks. These figures apply to what is called 'mechanical' pain, which means pain that is made worse by movements and is relieved by rest. Even if you still have some residual pain after six weeks, you should continue to increase your activity level progressively even if it hurts a little since this will speed up your recovery – do not be afraid that you are making yourself worse.

Chronic pain is pain that lasts more than three months. However, it is likely that your doctor will arrange for further investigations or a specialist referral if there has been no improvement after about four to six weeks.

29 What treatment
is a doctor likely to
be able to offer for
an attack of acute
back pain?

The doctor's first aim will be to make a
diagnosis, which is based partly on your
description of the pain and how it arose, and partly
on what she finds when she examines you. For
example, there may be local tenderness in your
back, it may hurt when she raises your leg from the
couch, or the reflexes in your legs may be absent. It
may not be possible for her to say exactly what
caused your problem (the cause of back pain often
remains unknown), but she will at least be able to
exclude the likelihood of a serious cause for the pain,
such as infection or cancer, that would require
urgent investigation and treatment.

As regards treatment, she will probably prescribe
medication of some kind. The main purpose of this is
to relieve pain. At one time aspirin was the chief
standby, but there is now a wide range of aspirin-like
drugs called non-steroidal anti-inflammatory drugs
(NSAIDs). These do tend to reduce inflammation, as
the name suggests, although it is uncertain how
important this aspect is in the case of back pain.
Their main action is to relieve the pain, which they
do effectively for many people. Unfortunately, they
also have side-effects, mainly on the stomach, where
they can cause ulceration, especially if taken without
food or milk. For this reason some patients are
unable to tolerate them and they are unsuitable for

people with a history of ulcers of the stomach or duodenum. It is sometimes possible for your doctor to give another medicine to reduce acid secretion in the stomach and so reduce the likelihood of adverse effects from the NSAID.

Although all the different NSAIDs work in much the same way, there are individual differences in people's reactions, so if you find that the medicine you have been prescribed isn't helping much you should tell your doctor, who may wish to change to a different member of the group.

Sometimes a different class of drug may be prescribed, such as a member of the opium family. There is no danger of addiction occurring in such cases because the drugs are used for short periods only and because addiction does not usually occur if these drugs are given for pain.

Muscle relaxant drugs are also sometimes given and can help, but they tend to cause drowsiness and this often limits their usefulness.

Some doctors may provide non-drug treatments in addition. These could include the loan of a TENS machine (*see Q 30*). These days an increasing number of doctors either practise some form of complementary medicine themselves (acupuncture, *see Q 88*, osteopathy, *see Q 86*) or have someone working with them who uses these treatments. Even if they do not, many are sympathetic to this kind of approach and will give advice about where to go for such treatment.

30 What exactly does a TENS machine do? Is it worth trying and is it safe?

TENS (transcutaneous electrical nerve stimulation) is a form of treatment that consists of the application of a small electrical current to the nerves through the skin. It is an effective form of pain relief for many people and is very safe. The only precautions to note are that it should not be used by heart patients with a fitted pacemaker and it should not be applied around the eyes or on the front part of the neck. Although it does not cure the underlying problem, it can give pain relief while natural recovery is taking place.

It is still not entirely clear how TENS works but there is no doubt that it does work, and not just by means of suggestion. Part of the explanation is probably to be found in the 'gate theory' of pain. It is thought that there are gates or filters in the spinal cord that can block the transmission of pain messages to the brain. These gates are closed when impulses arrive via certain types of nerve fibre (the large diameter nerve fibres), and it is these large fibres that are stimulated by the TENS machine, thus closing the gates, even if only for a short period.

The apparatus consists of a small box, about the size of a tape disk. It has either two or four leads coming from it, with conducting pads (electrodes) attached to them. These pads are fixed to the patient's body and the machine is turned on. The

Electrodes

This illustration shows a typical TENS machine with heads and electrodes. See page 54–55 for the placing of the electrodes.

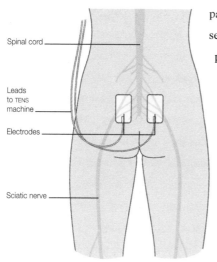

Spinal cord

Leads
to TENS
machine

Electrodes

Sciatic nerve

The electrodes are placed on the body near to the origin of pain. The electrodes shown here have been placed to help relieve sciatic pain.

patient should experience a tingling sensation, usually at only one of the pads, and relief of pain generally follows rapidly. The relief continues as long as the machine is switched on but usually fades fairly rapidly when stimulation stops; however, some people experience prolonged relief after stimulation. The machines are battery operated and therefore the total current applied is well within the safety zone; there is no possibility of serious electric shock. The worst adverse effect that may be encountered is skin irritation from the pads; this can sometimes be alleviated by shifting their position slightly. The machine can be left on for many hours at a time and you can walk about quite freely while it is working.

All TENS machines have at least two controls. One regulates the strength of the stimulation. You must be able to feel this; unless you feel the tingling sensation, you will not get relief. The strength of stimulation, however should not be so great as to be unpleasant. The other control regulates the frequency of stimulation, which is usually in the 50–120 Hz range. (This means that the number of individual electrical pulses per second is in this range.) The ideal frequency varies somewhat from patient to patient but is usually about 80 Hz.

The most critical decision when using a TENS machine is where to place the conducting pads. Usually the best plan is to place them on either side of the painful area; so for pain in the lower back you might place them on either side of the spine. Another idea is to place them over a major nerve going to the area, so that for sciatica you might place them over the sciatic nerve itself, (see *illustration*). There are also other possibilities, which are usually described in the instruction leaflet enclosed with the machine, and it is worth experimenting with a number of positions to find the one that works best for you. You should also try different frequency settings, because although 80 Hz is best for many people there are some who will respond better to faster or slower settings. Some machines provide a 'burst' facility, which gives short trains of current impulse, usually twice a second. This may be more effective for certain patients.

TENS machines are widely available nowadays, from mail order firms and elsewhere, and they have come down considerably in price. The more expensive versions have a wider range of settings (for example, some allow for more complex patterns of stimulation or for control of the actual duration of each pulse – the pulse width) but these refinements are not essential for most patients.

As with all self-help treatments, you should apply common sense and if the machine does not seem to be helping, you should consult your doctor for advice.

31 I'm suffering from severe back pain and have been told to stay in bed. How long should I stay there and when can I go back to work?

At one time it was thought to be beneficial to rest in bed to speed up the recovery process from back pain, but today the emphasis is more on early mobilization. Of course, as long as you experience a lot of pain when moving about you will have to stay in bed, but you should get up as soon as you reasonably can. Lying in bed causes your muscles to become weaker, which is undesirable, and prolonged rest leads to weakness of the bones due to osteoporosis. Moreover, as you move about, your muscles and joints keep sending information into your brain and spinal cord. This information competes with the pain information coming from your back and tends to reduce its impact. Hence, normal activities may tend to lessen the amount of pain you feel.

As regards work, you should return to it as soon as you feel that you can – there is no need to wait until the last vestiges of pain have disappeared. Naturally you should use common sense and if your work involves any heavy lifting or other physical activity that may have contributed to the problem in the first place, you may have to start by undertaking lighter duties.

I have seen various kinds of creams and ointments that you are supposed to rub in to relieve backache. Do these work?

Most of these substances, known as embrocations, act by causing local irritation of the skin, which may be associated with local flushing and a feeling of warmth. It is implausible to claim that this can have any effect on disorders affecting bones, joints, or muscles, but they may give some pain relief, partly by psychological means and partly by what is called counter-irritation: the local irritation caused by the embrocation stimulates the nervous system and this can reduce the amount of pain you feel by competing for your attention.

My doctor prescribed some anti-inflammatory tablets but my wife says the same thing is available as a cream to rub into my back. Wouldn't that be better?

It may seem logical to apply anti-inflammatory creams to the painful area, but, as noted above for embrocations, you cannot really affect the bones and joints by applying substances to the skin. What happens when these creams are used is that the drug enters the bloodstream through the skin and then travels round the body to produce its effects, in much the same way that it would do if taken by mouth. You might expect that applying it to the skin would reduce the chances of it causing side-effects on the stomach, but this isn't the case; the risk of stomach damage is still present. So it makes little difference by which method the medication is given.

Long-term
Problems

Some people suffer from long-term or recurrent back pain. If this is happening to you, you may be wondering what causes the pain and whether you will ever be free from it. This section explains some of the causes of chronic back pain and describes options for treatment, including orthodox and complementary therapies, and how you can ease the pain for yourself.

34 I have low back pain on and off for much of the time and now I have been diagnosed as suffering from chronic low back pain. This sounds rather alarming. What does it mean for the future and how should I cope with future attacks of low back pain?

Chronic low back pain is defined as back pain lasting for more than 12 weeks. The term does not imply anything about the cause of the pain and there are many possibilities. The outlook will depend on which applies in your case.

Many specialists nowadays think that chronic back pain is really a different disorder from acute back pain. Most people with acute pain recover quite quickly, so why do some people have pain that lasts, on and off, for months or years? A lot of time may be spent looking for a specific cause, but as a rule none is found, or the problem may not lie in the back at all. Studies have shown that back pain is strongly associated with chest and heart disease, smoking, depression and adverse social conditions.

The link with chest and heart disease and smoking is probably explained by a narrowing of the arteries that supply the spine. This can damage the intervertebral discs. The link with depression and adverse social conditions should not be taken to mean that the pain is all in the mind or unreal. Pain pathways in the spine and brain are known to be affected by people's moods and other psychological factors. Back pain seems to be one of the kinds of pain that is particularly liable to occur in this way.

Therefore it is likely that, for many patients, chronic back pain is not just acute back pain that goes on for a long time, but is a different problem.

This means that you should think about your back pain in the context of your lifestyle and you may need to discuss all this with your doctor. It may be possible for you to change some of your circumstances (if you smoke, you could give it up), but you may have to accept that you will have further episodes of acute back pain from time to time.

Don't think of yourself as an invalid. If you do have a flare-up of back pain, it is treated in the same way as in an initial attack: mainly by using drugs from the NSAID family (see Q 29). TENS (see Q 30) may also be helpful. If possible avoid bed rest and in any case do not stay in bed for more than two days at a time. Exercise (see Q 12) can work to condition the muscles of your abdomen and spine, which helps to prevent recurrence of pain, or you could consider hydrotherapy (see Q 36).

CHRONIC OR RECURRENT BACK PAIN

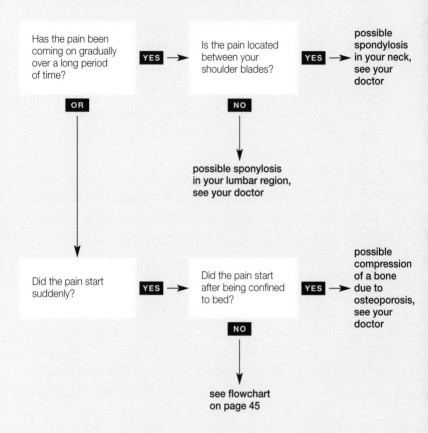

Has the pain been coming on gradually over a long period of time? **YES** → Is the pain located between your shoulder blades? **YES** → possible spondylosis in your neck, see your doctor

OR

NO ↓

possible sponylosis in your lumbar region, see your doctor

Did the pain start suddenly? **YES** → Did the pain start after being confined to bed? **YES** → possible compression of a bone due to osteoporosis, see your doctor

NO ↓

see flowchart on page 45

You should always consult your doctor when suffering from unexplained back pain.

35 What is a
consultation with
a specialist like?
Will I have to have
an operation?

The specialist may be a neurologist, a neurosurgeon or an orthopaedic surgeon. In describing your symptoms, try to confine yourself to what you actually feel and avoid offering a diagnosis. Often, patients use expressions that mean one thing to them and a different thing to the doctor.

Quality of pain is often hard to describe: words such as throbbing, burning, sharp and dull seem to mean different things to different people. But notice particularly which parts of your back or leg are affected and the influence of changes in your position or activities (standing, sitting, lying down; walking, driving, coughing, sneezing). Mention any changes in sensation, or pins and needles. Noting the times at which the pain is worse can sometimes be helpful. Also, be prepared to give a clear account of how and when the symptoms first appeared, any changes that have occurred subsequently and the relief, or absence of relief, that you have experienced from any medicines or other prescribed treatment.

It is by no means certain that a specialist will recommend surgery – only about 1 in 10,000 people with back problems ever has surgery. If this is not advised in your case, the specialist may suggest a range of other options, including physiotherapy, hydrotherapy (see Q 36), acupuncture (see Q 88), or use of a TENS machine (see Q 30).

36 What is hydrotherapy and is it beneficial in chronic low back pain?

Hydrotherapy is the performance of specially designed exercises in a pool, under the supervision of a physiotherapist. When you do these exercises in the water, your weight is supported, which reduces the strain on the joints and allows movements to occur which would not necessarily be possible on land. Also, the warmth of the water relaxes your muscles and tends to relieve the pain in your spine, so that you can move more freely. If you find getting into the pool difficult, a hoist is usually available. There is no need to be able to swim in order to derive benefit from hydrotherapy.

Studies have shown that three weeks of hydrotherapy reduces the intensity and duration of back pain, reduces the amount of pain-relieving drugs needed and improves spinal mobility. The benefits are sustained for many months after treatment, so this form of treatment is well worth considering if it is available to you.

I was sent for an x-ray of my neck and my doctor said it showed a bit of arthritis but nothing to worry about. However, I now feel very worried. I'm only 45; will I be crippled by the time I'm 60?

It is really not a good idea to use the word 'arthritis' in this connection, because many people take this to mean a progressive disease that will eventually lead to complete incapacity. The changes commonly seen on X-ray are what is know as spondylosis (not spondylitis). The expression 'wear and tear' is an improvement.

Spondylosis refers to the changes that occur in the spine with age. They are of various kinds. The small joints between the vertebrae (facet joints) are liable to develop osteoarthritis, a non-inflammatory disease. This may give rise to pain in its own right or may cause the joints to swell and press on the nerve roots, giving rise to secondary (referred) pain.

The intervertebral discs may also deteriorate with age and X-rays may show that one or more of your discs is thinner than normal (so-called worn discs). Narrowing of the space between the vertebrae may then compress the nerve roots. Other changes of age-related degeneration may also be evident on X-ray. The most dramatic are outgrowths of bone called osteophytes, but these often cause no symptoms and merely limit movement of the spine.

The neck is vulnerable because it is the most mobile part of the spine. A specialist will usually ask for blood tests and an X-ray. CT or MRI scans are usually only done if there is the prospect of surgery.

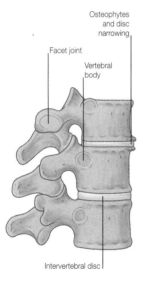

Osteophytes and disc narrowing

Facet joint

Vertebral body

Intervertebral disc

When suffering from osteoarthritis joints become deformed, in this case causing pain in the lumbar region of the spine.

I find I get several
different types of
pain in my back.
Some are sharp and
stabbing, others are
dull and diffuse. Why
does this happen and
what does it mean?

In broad terms there are three types of
'mechanical' spinal pain, although they often
occur together. The first kind arises from what is
called the 'motion segment': that is, a pair of
vertebrae and the joints between them (the
intervertebral disc and the two facet joints, one on
each side). The muscles and ligaments in the region
also form part of the motion segment. Any of these
structures can give rise to pain.

Pain arising from a motion segment has a deep,
dull, aching quality and it may radiate to other areas.
For example, pain originating in the neck region may
radiate to the eye, the chest wall or the elbow; pain
in the thoracic region may be felt in the front of the
body, giving rise to a mistaken diagnosis of
abdominal pain; and in the lumbar region there may
be radiation to the groin, lower abdomen and leg,
even as far down as the foot. Changes in blood flow
in these areas can occur as well and patients may
experience local sweating or other effects.

The second type of pain affects the superficial
tissues, including the skin, the superficial ligaments
and muscles and tips of the spines of the vertebrae.
This kind of pain can be accurately localized, unlike
the first type. One example is 'kissing spines', which
is felt in the middle of the back when it is straight
but disappears when the patient bends forward.

The third type of pain occurs when a spinal nerve is compressed as it emerges between two adjacent vertebrae. This pain is sharp and tends to resemble that of an electric shock. It is similar to what we experience when we hit the 'funny bone' (ulnar nerve) in the elbow. There is often some loss of sensation, or an increase in sensitivity, in the area of skin supplied by the affected nerve and this helps your doctor to identify the spinal segment involved. If the motor function of the nerve is affected there may ultimately be wasting of the muscles supplied by the nerve and the limb reflexes may be lost.

There are wide variations in what makes all these pains (but especially the first kind) better or worse. Some people feel worse standing, others worse sitting or even lying down. Coughing and sneezing may also exacerbate the pain. Bending may be all but impossible as a result of muscle spasm, so that you cannot put your shoes and socks on, or your back may move quite freely. Your back itself may be very sensitive to pressure, sometimes just over one or two vertebrae, or there may be no local tenderness at all.

Back pain is often worse at night. Most patients with osteoarthritis find that they stiffen up with prolonged rest and this may be another aggravating factor. Changes in position during sleep may also play a part. However, if pain is always much worse at night, regardless of your position, and especially if you are aged over 50, you should see your doctor to check that there is no serious underlying cause.

Q

39 My son, aged 18,
has been told he
has ankylosing
spondylitis. What
is this and can it
be cured?

Ankylosing spondylitis is an inflammation of the spine. It affects men much more than women and there is a tendency for it to run in families. It causes pain and stiffness; in advanced cases the sufferer cannot move his spine at all. Many people, however, have the disease in much milder form. It nearly always begins in the sacroiliac joints of the pelvis and spreads upwards to involve the small (facet) joints of the spine at higher levels.

It usually begins with pain in the low back or hip and sciatic regions and a lot of morning stiffness. Later the pain begins to occur at higher levels in the spine. Characteristically, the severity of the disease varies, with aching and stiffness lasting for a few days or weeks at a time and then subsiding almost completely. Sometimes joints outside the spine, such as the knees, are involved. The disease may become arrested at any stage and progress no further.

It is diagnosed by x-rays and blood tests. There is no cure, but the symptoms can be helped to a considerable extent by treatment. Pain-relieving medicines are prescribed as needed and physiotherapy is also helpful. Patients are encouraged to perform daily exercises to maintain the mobility of their spines as much as possible. It is also worth considering acupuncture (*see Q 88*), which can relieve the pain for some patients.

40

I had an x-ray of my back and it showed spondylolisthesis of the fifth lumbar vertebra. What does this mean and what can be done about it?

Spondylolisthesis is a slipping forward of a vertebra. It may occur as the result of a single severe injury or many minor injuries; it can also arise during growth. There may be no symptoms and the abnormality may be found in the course of an x-ray or CT scan. If symptoms do occur, they often take the form of low back pain radiating into the legs and there may be tenderness when the affected vertebra is pressed. If the degree of slippage is considerable, there can be a 'step' in the back where the upper part has moved forwards on the lower part; in these cases the patient's trunk may be shortened and the abdomen may project forwards. There may then be pressure on the nerve roots.

A somewhat different form occurs in older women and some men. The fourth lumbar vertebra moves back and forth more than normal in these people and they suffer from low back pain, sometimes for years and get pain in the buttocks and thighs. The pain is felt on both sides of the back, though one side is usually worse than the other. Standing for long periods is painful but so is sitting still in one position.

Spondylolisthesis is diagnosed by CT scans or x-ray taken from the side. If there are no symptoms, or they are easily controlled, there is no need to do anything. If symptoms are severe, the slipping vertebra can be fixed in place by an operation.

Q

41

My mother, aged 70, has been told she has a crush fracture of a thoracic vertebra. She hasn't had an accident, so how did this happen?

In your mother's case this is almost certainly due to osteoporosis. The loss of calcium and bone tissue results in vertebrae becoming weaker and liable to break. It does not take much to cause this: a misstep while coming downstairs or even a violent cough may be enough. A crush fracture of this kind causes pain in two ways. The fracture itself gives rise to pain and the loss of space between the vertebrae may press on nerve roots.

If a number of vertebrae collapse, as may happen over the years, the result is loss of height. This is why some elderly people become shorter as they age. X-rays taken from the side will show increased curvature of the spine and one or more vertebrae that have taken on a wedge-shaped appearance, with the narrow end towards the front.

Osteoporosis typically affects women after the menopause. It tends to get worse with age although it can be prevented to some extent by hormone replacement therapy (HRT). Regular exercise helps, as does taking calcium and vitamin D. Men can also suffer from osteoporosis.

In younger patients, crush fractures may occur after a bad fall, especially if the person lands on their buttocks. There is no specific treatment for fractures of this kind; patients are given pain-relieving drugs and the fracture is allowed to mend itself. A major exception is a fracture of the vertebra called the axis, at the top of the spinal column. If this is left alone the vertebra can slip and cause sudden death, so surgery is needed.

Injury to the coccyx at the bottom of the spine can give rise to severe pain. Some people suffer pain of the coccyx without having an injury; the cause of this pain is obscure and it can be difficult to relieve.

42 **My mother, aged 60, is suffering from osteoporosis and gets a lot of backache although x-rays do not show a fracture of the spine. Why is her osteoporosis painful?**

Osteoporosis can cause back pain by weakening vertebrae so that they collapse, as described in the previous question. In the absence of such collapse, osteoporosis is usually said not to be painful; however, some patients, mainly women, undoubtedly do suffer pain from the osteoporosis itself. The cause is unknown, but one theory is that it is due to distension of the veins in the bones.

For treatment, your mother could use ordinary pain-relieving tablets such as paracetamol. It could also be worth trying physical treatments such as TENS (*see Q 30*) or acupuncture (*see Q 88*), both of which can offer some relief.

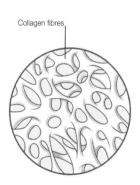

Collagen fibres

The top image shows a normal bone, the bottom illustration is of a bone affected by osteoporosis.

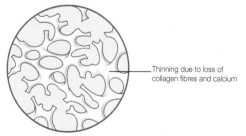

Thinning due to loss of collagen fibres and calcium

I suffered a whiplash injury a year ago when involved in a minor traffic accident and I still get neck pain. Why is this and can anything be done to relieve the pain?

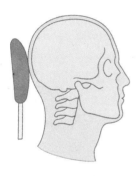

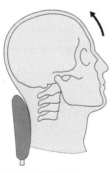

It is essential that car headrests are high enough to reach the back of the head. A head rest that reaches only as far as the neck is not merely useless but positively harmful.

Whiplash injuries typically occur when a vehicle is impacted from behind. The force of the blow moves the car and its occupants forwards, but their heads tend to remain in the same place, so there is a sudden flexing of their necks. This can be severe enough to dislocate the spine or cause vertebral fracture and surgery is needed in such cases. Crash victims should not be moved until the ambulance crew arrives, because a displacement of the fractured vertebra could cause paralysis or death.

In less severe impacts there is no definite fracture or dislocation but patients continue to suffer symptoms for long periods after the accident. Many different effects are reported, including pain in the neck, shoulders or arms and tingling sensations in these regions. There is much dispute among experts about what causes these symptoms. X-rays and scans of the neck often show few abnormalities and even when changes are seen it can be difficult to know whether they were present before the accident. Symptoms may continue for months or even years, especially in women.

Treatment of whiplash injury is similar to the treatment of neck pain of other types. Pain-relieving medication may be given. Other approaches include physiotherapy, osteopathy or chiropractic (*see Q 85 and Q 86*) and acupuncture (*see Q 88*).

44 **I had a routine chest x-ray that showed that I have an extra rib on the right side. Is this anything to worry about?**

The extra rib is called a cervical rib, because it arises from the lowest (seventh) neck vertebra, which does not normally carry a rib. It may be present on one side or on both. Such extra ribs may cause symptoms by pressing on the blood vessels in the neck. Sometimes, instead of a rib, there is a band of tissue connecting the seventh cervical vertebra and the first rib; this can press on nerve roots and cause weakness of the small muscles in the hand and impaired sensation in the little and ring fingers. If symptoms are troublesome, the abnormal rib or band of tissue can be removed surgically, but if there are no symptoms there is no need to do anything about them.

Some patients have very similar symptoms although no cervical rib is actually present. The cause of the problem in such cases can often not be diagnosed and treatment must then be directed purely at the symptoms. ●

Extra ribs, known as cervical ribs, can occur on almost any vertebrae, here we can see one on the seventh cervical vertebra.

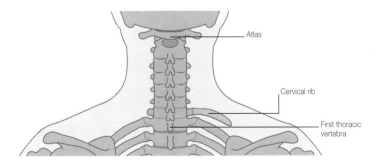

Atlas

Cervical rib

First thoracic vertebra

Q

45 My uncle, aged 50, had a sudden onset of pain in his shoulder and later his arm became weak. The doctor thought this was due to arthritis of his neck, but then he saw a specialist who said it was acute brachial neuritis. Can you explain what this is?

Acute brachial neuritis is often confused with shoulder and arm pain associated with neck problems, but the cause is different. The symptoms are as you describe in your uncle's case. Initially there is severe pain in the shoulder or shoulder blade (scapula) and some days or weeks later this is followed by weakness of the shoulder and upper arm. The cause is unknown, although there sometimes seems to be a link with previous infection or immunization. It is important to make the diagnosis of this disorder, because most patients recover slowly but completely over the next two to three years, whereas if the symptoms are due to pressure on nerve roots in the neck, spontaneous recovery is then unlikely.

Q

46 My husband, aged 70, gets pain in both thighs when he walks any distance. Bending over relieves it. He has been diagnosed as suffering from neurogenic claudication. What is this?

Claudication, which means 'limping', refers to disturbance of walking caused by insufficient blood supply. The most frequent form is intermittent claudication, in which impairment of blood supply to the legs causes the sufferer to feel pain in the legs when walking, so that he is forced to stop. This form of claudication is typically due to arterial disease.

The disorder that your husband is suffering from is different and less common – due to impairment of circulation in the nerve roots in the lumbar region. It tends to come on in later life because of spondylosis (*see Q 37*), but usually the patient already had a spinal canal that was narrower than normal. The symptoms are unpleasant feelings in the legs, which patients describe as numbness, coldness, burning or cramps, and which can amount to actual pain.

The characteristic feature is that the symptoms come on when the patient stands upright, because in this position the spinal canal becomes narrower; they are relieved by bending forwards. This gives one way of distinguishing between neurogenic claudication and intermittent claudication. Patients with the latter prefer to walk downhill, because this is less demanding, whereas patients with neurogenic claudication prefer to walk uphill, because then they bend forwards slightly. For the same reason, patients with neurogenic claudication may find that they can cycle without problems even though they find walking difficult. This is because in cycling they tend to bend forwards to lean on the handlebars.

Stooped posture caused by spinal stenosis. The leaning forward may not be caused by the disease itself but eases the pain in this disorder.

Good posture

Undesirable

Cycling in the inclined posture improves the circulation in the nerve roots.

47 Can a psychological problem like depression cause backache?

The boundary between psychological and physical illness is not a sharp one and it is not uncommon for depression to manifest itself in physical as well as psychological pain, especially back pain. This is often called somatization, which means the expression of psychological problems in a physical form. It is important to understand that the pain such patients experience is perfectly real, although it is not due to local physical changes in the back. It is due to changes in the way in which the brain processes incoming signals from the body. Modern techniques for electronically imaging the brain show that the pain in these cases is accompanied by actual visible changes in the way the brain is functioning.

It is not helpful to continue to search for a physical explanation for back pain in such cases, because studies have repeatedly shown that if nothing physically abnormal is found initially, it is extremely unlikely that a serious physical problem will be found later. Treatment directed at the back in such cases will be useless. What is needed is treatment of the depression: antidepressant drugs, with psychotherapy or other counselling approaches designed to alter the patient's attitude and expectations. Both methods may, of course, be used at the same time.

Even when there is a definite physical cause for a patient's back pain, psychological factors may still need to be taken into account. Any form of long-lasting pain is likely to cause some degree of depression and this may need treatment in its own right. For some people, continuing back pain may serve a psychological purpose, for example by giving them an excuse to avoid an unpalatable situation at work or elsewhere. In such cases, backache may last for years because it serves a purpose in the patient's life. None of this means that psychological considerations impose an imaginary layer on the physical illness, but rather that the nervous system always modifies the amount of pain that is perceived in any situation.

Another point to note is that many of the drugs that are commonly used to treat depression have a pain-relieving effect quite independently of their antidepressant effect and are sometimes prescribed on that basis. So if you are asked to take an antidepressant to relieve your pain, this does not necessarily mean that the doctor thinks you are feeling depressed.

Q

48

48 I have had every imaginable treatment for my backache, including surgery, osteopathy and acupuncture, but nothing has helped. Now my doctor has referred me to a pain clinic. What will they do there that hasn't been tried already?

Many hospitals nowadays have a pain clinic, which is usually headed by an anaesthetist. Such clinics have existed for some 50 years. The idea is to provide a single site at which a wide range of pain treatments is on offer.

When you arrive you will be seen by a doctor, who will then be responsible for your care and follow-up. This doctor will take a history and arrange for any necessary tests, after which he or she will decide which of various treatments is most likely to be useful for you. These may include different combinations of drugs, the use of electrical stimulation, nerve blocks and other techniques. The advantage of a pain clinic is that the staff have wide experience of these methods and can bring an exceptional degree of expertise to bear.

Another aspect of pain control that is often particularly important for back sufferers is education about what does and does not make their pain worse. Patients need to understand that keeping their back mobile will actually diminish the pain they feel and prevent recurrence. Physiotherapists in pain clinics are able to explain this.

Pain clinics can show patients different ways of coping with the pain that make their lives worth living – in spite of the pain. This approach is called cognitive therapy (*see Q 49*).

Although much can be done to help patients with back pain, there will always be a few unfortunates who fail to respond to any kind of treatment. In such cases it is generally best to find ways of coming to terms with the situation – not in the rather negative context of 'learning to live with the pain', but in finding ways to accommodate the pain without being completely dominated by it. This is what cognitive therapy is designed to do.

There are several components in this approach. One is that it may be possible to alter your perception of pain by changing your attitude to the situation. Another is that people who try to 'protect' their backs by doing as little as possible may be making the situation worse. The muscles and joints deteriorate if they are not used and if the muscles are weak the back is more prone to injury.

At some pain clinics, patients are taught specific mental techniques for coping with pain and minimizing its effects. There is some evidence that these methods can set up a 'virtuous circle' of cause and effect – the pain does not necessarily go away but patients are enabled to live fuller lives in spite of the pain. Treatment is carried out either individually or in groups. Psychological 'rewards' for pain, such as sympathy, are removed. Techniques for distracting your attention from the pain are taught.

Q

50 Can changing my diet relieve my back pain?

There is not much evidence that diet has a lot of effect on back pain. The main exception to this is in cases of rheumatoid arthritis, which may cause back pain if it affects the spine. Diet can influence the severity of this disease, although it only helps about five per cent of patients. This group finds that they can reduce the severity of their disease by avoiding certain foods.

It is possible that there are more patients who would be helped by avoiding particular foods, but it is very difficult to find this out for yourself. You would need to spend several months under professional supervision. You would first be given an 'elemental' diet containing the most essential nutrients and then foods would be introduced one at a time to see which made you worse. It is quite unsafe to try to do this on your own, because you might develop severe nutritional deficiencies.

At least one study, from Scandinavia, has found that meat and fish make some patients with rheumatoid arthritis worse and that following a vegetarian diet (with plenty of dairy products) is helpful. This could be worth considering for anyone who wishes to make the experiment.

Do not be tempted by advertisements offering tests of hair or blood to detect deficiencies or allergies and treat arthritis – they are quite unreliable.

I have been referred to the occupational therapy department of my local hospital for my neck pain. I do not understand why I'm being sent there. Can you explain?

Occupational therapy specialists can contribute to the management of back and neck pain in at least two ways. In many departments they offer various forms of relaxation. Some of these are quite simple, while others, such as autogenic training (*see* Q 97), are more elaborate. There is no need to feel defensive about this or to think that this means your pain is not real. On the contrary, tension of this kind is recognized as a major cause of pain.

It is not clear why some people respond to pressure in this way. One theory is that tensing of the neck muscles is a reflex action in response to a threat, which has been programmed into our nervous system during evolution. It can be difficult to treat patients who react in this way, hence the need to try some form of relaxation. Another possibility is regular physical exercise, which has a natural tranquillizing effect.

The second way in which occupational therapists can help back sufferers is with advice on everyday living. Common activities such as dressing, hair washing, making-up and shaving can pose difficulties, but there are often simple solutions. A long-handled shoe horn will help with tight shoes, while leaning forwards across a basin to peer at yourself in a mirror places undesirable strain on your lower back; it is better to sit down.

52 A friend of mine has had a lot of back pain, which she has been told is due to arachnoiditis. What is this?

Like the brain, of which it is an extension, the spinal cord is covered with three layers of membrane. The outermost is a tough layer called the dura mater. The innermost is a very fine layer called the pia mater. Between the two is a delicate membrane called the arachnoid mater (arachnoid refers to its texture, which is supposed to resemble that of a spider's web). 'Arachnoiditis' means an inflammation of the arachnoid mater. In the past it most often arose as a side-effect of dyes that were injected into the spinal cord to make it visible in x-rays. Modern imaging methods, such as MRI, do not use such dyes, although they are at times used as part of a CT scan. Arachnoiditis still occurs at times, sometimes after operations on the back, sometimes after infections and sometimes for unknown reasons. It does not always produce symptoms but it can give rise to persistent back pain that is difficult to treat. It may clear up spontaneously without treatment.

White matter

Spinal cord

Grey matter

Dura mater

Pia mater

Arachnoid mater

A healthy spinal cord with its protective coverings.

53

My father, aged 70, has pain in his lower back which he is told is due to Paget's disease of bone. I have never heard of this; what is it?

Paget's disease of bone is a disorder in which areas of bone are destroyed, and then are spontaneously repaired. The cause is unknown; one idea is that it is due to a virus. It may affect any of the bones, including the vertebrae, but is particularly common in the lumbar spine and the pelvis. In some cases the skull may also be affected. There may be no symptoms at all, in which case the disease is discovered by chance when an X-ray is taken for other reasons which shows the bone distortions. Other patients can have persistent pain, which is usually dull but may be shooting or knife-like. It occurs most often in the lower back and may radiate into the buttocks or legs. The pain may be due to the Paget's disease itself or can be caused by consequent distortions of the bone structure.

Most patients require no treatment since they have no symptoms. However, if pain is present, then pain-relieving drugs (NSAIDs, see Q 29) are used. There are also drugs, called biphosphonates, which are moderately effective in reducing the rate at which the bone changes occur; this helps to alleviate the patient's symptoms although relief is not usually complete. The biphosphonates have largely replaced the older drug, calcitonin, which was previously used for this purpose, though calcitonin is still useful for some patients.

54 I get low back pain
quite often and I'm
always afraid that it
may be cancer.
Should I worry?

It is true that cancer may give rise to bone pain,
including pain in the back. However, this is not a
common cause of back pain. If you are getting fairly
frequent attacks of low back pain over a long period
of time, it is almost certainly not due to cancer.
However, if you are worried you should see your
doctor, who can arrange for X-rays or other tests to
set your mind at rest.

There are certain circumstances in which caution
is needed. A first attack of back pain in someone
over 50 should be investigated, because the older
you are the more likely cancer is to develop. Severe
pain at night or pain that is much worse when you
lie down require further investigation. If you have
already had treatment for cancer, your doctor or
specialist will want to make sure that any recent
onset of backache is not due to a recurrence. This
can usually be done with a plain X-ray, although
sometimes a bone scan is performed. In this, a short-
lived radioisotope is injected and its rate of uptake in
the spine is measured. The exposure to radiation
involved is negligible.

55 My father, aged 74, finds he gets dizzy when he turns his head. A physiotherapist told him this is due to wear and tear in his neck. How does this cause dizziness?

Your physiotherapist's explanation is probably correct. The reason is that your father is suffering from osteoarthritis of the small facet joints in his neck. The sensory nerves that supply these joints are closely connected with the nerves that supply the balance organ in the inner ear. When we move our heads, information goes from the joints in the neck to the central nervous system to let it know what is happening; in this way we can compensate automatically for neck movements and not become unbalanced when, for example, we turn the head to look at something while walking.

In your father's case the disease of his facet joints is causing abnormal messages to go to his balance mechanism, upsetting his sense of balance. There are various ways of reducing this problem to manageable proportions. One is to fit him with a collar to restrict his neck movements. Another is by means of retraining exercises – your physiotherapist can advise about this. Yet another approach is acupuncture to the neck (*see Q 88*).

A different explanation might be that osteoarthritis of the neck is causing pressure on the vertebral artery, but this would be more likely to cause brief unconsciousness than dizziness. Other possible causes for giddiness (vertigo) are unrelated to the neck and would need special investigation.

56 I get numb hands. At first I was told this was due to carpal tunnel syndrome but now they say it is coming from my neck. Can you explain this?

Carpal tunnel syndrome is due to pressure on the median nerve at the wrist. It is generally due to a band that stretches across the front of the wrist, called the flexor retinaculum, which serves to keep the tendons in place when you bend your wrist. It particularly affects women at about the time of the menopause and can also occur during pregnancy. Initially it gives rise to pain in the hand and wrist, which is usually most noticeable in the mornings. Occasionally the pain extends up the arm towards the elbow. There may also be numbness of the thumb and first two fingers. If it gets worse, weakness of the muscles in the hand (especially those of the thumb) may occur and they become wasted. Usual treatment is surgery to relieve the pressure, and sometimes injections of corticosteroid under the ligament thus avoiding the need for an operation.

Pressure on the nerve roots in the neck can give rise to a rather similar picture, with pain, tingling, loss of sensation and weakness in the arms and hands, but the treatment naturally needs to be different because the underlying mechanism is different. In order to distinguish between these causes, nerve conduction studies are performed. In these, the rate at which impulses are transmitted along the nerves is measured and this allows the surgeon to know at which level the obstruction lies. ●

Carpel tunnel syndrome affects the thumb and first two fingers. The rest of the hand feels completely normal.

Area affected

What is referred pain and why does it cause backache?

Referred pain is pain felt at some distance away from the site of trouble. Disease of any internal organ may produce pain felt mainly in the back. This is a result of the way the nerve supply to the organs is distributed from the spinal cord, in what is termed a segmental pattern. Disease in the pelvic organs is generally referred to the lower back (sacrum), disease in the lower abdomen to the lumbar region and disease in the upper abdomen to the lower thoracic (chest) region. As a rule there are no local signs in the back: the spine moves freely.

The pain of ulcers in the stomach and duodenum is usually felt in the upper abdomen, but sometimes it causes back pain. An indication of this might be pain that is worse after you eat an orange or drink alcohol or coffee; it may be relieved by drinking milk or taking antacid medicine. Disease of the pancreas and of the lower bowel (colon) can refer pain to the back, instead of or as well as pain in the abdomen. Gall bladder pain may be referred to the back and may be brought on by eating fatty food.

Kidney disease causes pain referred to the back, as does disease of the prostate gland at the base of the bladder. In these cases the pain is often felt in the lower part of the back (sacrum). If a patient on anticoagulant medicine experiences sudden back pain, a doctor should be seen immediately.

My physiotherapist tells me that the pain in my neck and shoulders is due to trigger points. What are trigger points and what causes them?

Trigger points are areas in muscles that hurt when pressed and from which pain, called referred pain may radiate to other areas.

We do not know exactly what trigger points are. They seem to be areas that are active electrically and where the blood flow is probably increased. However, if they are removed surgically, nothing abnormal is found. Trigger points can disappear quite rapidly, for example in response to simple pressure with the fingers, although they come back later. Therapists detect them in various ways but mainly by examining the patient with their fingers. The trigger points are tender, sometimes acutely so, and may be associated with bands of taut muscle fibres. When pressed or injected, the trigger point may give rise to a localized muscle twitch.

Trigger points do not always cause symptoms, in which case they are called latent. At times, however, one or more trigger points may become active and give rise to pain, which may be situated quite a long way from the actual trigger point. For example, trigger points in the lower back may cause pain in the legs and so on. In some cases they give rise to weakness or changes in circulation rather than to actual pain.

Trigger points are often found in muscles such as the trapezius and can cause severe pain.

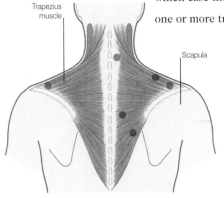

Trapezius muscle

Scapula

Various factors can cause trigger points to change from latent to active: these include sudden strains, undue exertion and emotional considerations. Once they have become activated they can persist for long periods – months or even years.

Treatment consists of deactivating the trigger point; this may be done in a number of ways. One is by spraying the affected muscle with a cooling spray and then stretching it. Another is by injecting local anaesthetic or other medication into the spot, or through acupuncture (see Q 88).

Active trigger points are more common in women than men. They tend to increase in number up to middle life and then begin to decline. ●

59 I have been told I have gout and that this is the cause of my back pain. I thought gout affected your big toe. Can you explain this?

Gout is due to a high level of uric acid in the blood. This can cause acute attacks of painful swollen joints, classically in the big toe. However, other joints, including those of the spine, may be affected and there may be severe pain in the muscles of the back, so that it is almost impossible to move. Gouty patients are particularly susceptible to painful osteoarthritis of the spine.

Acute attacks of gout are usually treated with anti-inflammatory drugs. Drugs are now available to reduce the level of uric acid in the blood and this should prevent acute attacks from occurring and also prevent damage to the spine. ●

60 What is the difference between rheumatoid arthritis and osteoarthritis?

Rheumatoid arthritis is a disease producing generalized inflammation throughout the body. It can affect many organs, although principally the joints. It is mainly a disease of small joints (fingers, hands, wrists) although any joint may be affected, including those of the spine. Women are affected more than men. There are blood tests that can distinguish this form of arthritis fairly accurately from other kinds. Patients may lose weight, become anaemic or suffer from other unrelated problems.

Osteoarthritis is a disease that affects particular joints but is not widespread throughout the body and does not produce anaemia or other generalized effects. It is largely the result of 'wear and tear', so manual labourers and sportspeople are particularly at risk. The characteristic feature is that the articular cartilage, which is the 'bearing material' of the joint, is worn away, so that bone-to-bone contact occurs.

A form of osteoarthritis affects the hands of post-menopausal women, although in a different pattern from that usually seen in rheumatoid arthritis.

Osteoarthritis affects mainly the large weight-bearing joints, such as the hips and knees. It often affects the spine, either as part of the ageing process or from simple overuse. Symptoms generally begin in late middle age although they may appear earlier. Sometimes they are triggered by a sudden awkward movement or a fall.

I had a lot of pain in
my back after playing
tennis recently and I
was sent for x-ray.
This showed a stress
fracture of one of my
vertebrae. What is
this and what should
be done about it? Will
I be able to play
tennis again?

Any bone in the body may suffer a stress fracture
due to over-use or excessive muscle pull. In
the case of the spine, it usually affects the arch of the
vertebra, which is made up of two rather thin plates
of bone that meet in the midline and form part of the
spinal canal protecting the spinal cord. Stress
fractures in this region may occur when you increase
the amount of tennis you play or start playing again
after the winter. The pain is not necessarily very
severe and may be just a slight backache which
comes on after playing. However, if you ignore it, it
will become worse and you will feel it even when you
are not playing.

Diagnosis is by x-ray, but at an early stage a
fracture of this kind may be difficult or impossible to
see. To allow the fracture to heal you need to rest,
which means not playing tennis for the time being.
You will be able to return to it later, but you should
do so gradually. Your specialist will tell you when it is
safe for you to do so. It could be worth your while to
consult a sports physiotherapist, who can advise you
about exercises to build up strength and mobility in
your trunk.

I have been
prescribed
antidepressants
to relieve back
pain caused by
fibromyalgia. Can
you tell me what
this is and why
antidepressants
will help?

Fibromyalgia is a disorder characterized by widespread pain in the muscles of the back and shoulders; sometimes the hip girdle is painful too. The symptoms might suggest that the patient is suffering from polymyalgia rheumatica (*see Q 24*), however, there are no abnormal blood tests in fibromyalgia and there is no danger of temporal arteritis causing blindness. In fibromyalgia there are many trigger points (*see Q 58*), but these are not the same as the localized trigger points that are found in people who strain a particular group of muscles.

Fibromyalgia is more common in women, especially those of middle age. The cause is unknown. Most patients are somewhat depressed and there is usually sleep disturbance. Ordinary pain-relieving medicines, such as paracetamol, aspirin and NSAIDs, seldom seem to help much. Antidepressants are often used, partly because there may be associated depression, but partly because most of the drugs in this category also relieve pain in their own right; some may help with sleeping problems.

Physical treatments, such as physiotherapy, manipulation, massage and acupuncture, usually help (if at all) for only short periods. At present, fibromyalgia is a difficult disorder to treat and many patients have to find ways of coping with the pain, such as those taught in pain clinics.

Q3 My physiotherapist says I am suffering from postural pain. Can you tell me what this is?

The normal spine possesses several curves when looked at in side view, so that it tends to form an S shape. For example, the neck is slightly concave towards the back, the thoracic spine is convex towards the back and the lumbar spine is again concave towards the back. Convexity of the spine is called kyphosis, concavity is called lordosis. Both of these may be either normal in degree or exaggerated and abnormal. When they are unduly increased they may give rise to pain. Very tall people are liable to develop deformities of this kind, as are those who adopt unsuitable postures at work.

A change in posture is all but unavoidable in late pregnancy or if you are seriously overweight, but even some slim people adopt a slouched posture. This results in undesirable strain in the lumbar and thoracic regions and is an important cause of backache in younger patients. Older people have often adapted to this posture. However, the 'military' erect posture is also undesirable, causing aching in the lumbar region, neck and shoulders. Sitting posture is equally important and liable to abuse.

Correcting posture is difficult, because any alteration – even for the better – is likely to feel abnormal at first. Physiotherapists, osteopaths and Alexander technique practitioners (see Q 96) are specialists in postural re-education.

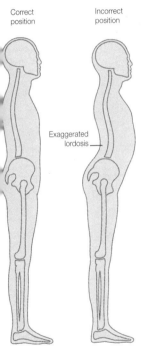

Correct position

Incorrect position

Exaggerated lordosis

Some people habitually stand in a slouched posture, with the stomach forward and a compensating 'hump' in the upper part of the thoracic spine.

Women &
Children

There are particular considerations in the diagnosis and treatment of back pain in women related to their reproductive system and to the stresses placed on the back during pregnancy. These are discussed in this section. We tend to think that back pain is a problem that primarily afflicts the middle-aged and elderly, and this is true. However, young people do suffer in this way and this section also offers information about the underlying factors of spinal disorders in children.

64 **My grandmother tells me that when she was young she was told that her backache was due to a misplaced womb and she had an operation for this. I too suffer from backache but my doctor does not think the same operation is likely to help. Why not?**

At one time it was fashionable to attribute pain in the lower back in women to minor displacements of the womb and operations were often carried out to correct these, but this is no longer believed today. However, it does remain true that more serious pelvic and gynaecological problems can cause back pain, especially when there is involvement of the ligaments that join the womb (uterus) to the sacrum. This can happen if the womb is prolapsed and pulls on the ligaments, in which case the pain may come on after you have been standing for several hours.

Another cause for pain in the lower back in women is endometriosis. This is a disease in which material similar to that which lines the womb is found in other places, such as the ligaments surrounding the womb, where it may give rise to pain. In such cases the pain is referred to the sacrum in the lower part of the back. Generally, this begins at some point in the days preceding the menstrual period and may even merge with the pain of the period making it almost impossible to diagnose. .

If prolapse is causing symptoms it should be treated, generally by operation, although sometimes a pessary is used to keep the womb in place. Endometriosis is treated with hormonal drugs. ●

35 **I only suffer from bad backache during my monthly periods. Why is this?**

Menstruation itself may be painful and the pain may be felt in the region of the sacrum in the lower back. This is a localized pain, cramping in nature, that may radiate down the legs.

If you have backache due to spondylosis (*see Q 37*) or other such back disorders you may find that the pain is more severe around the time of your periods. This is a general feature of pain for women, not just of back pain; many pains and other symptoms tend to get worse at these times. If pain at this time is a major problem you should see your doctor to make sure that there is no gynaecological cause that requires gynaecological treatment. ●

66 I'm seven months pregnant and I'm getting a lot of backache. Why is this and what can be done to help?

Backache is common in pregnancy. It affects the lower back and tends to radiate into one or both thighs. Partly this is due to the hormones that relax and soften ligaments in preparation for delivery, partly it is a result of the change in posture that inevitably occurs in the later months as the womb enlarges, obliging you to lean backwards when standing, causing an exaggerated lumbar lordosis.

Treatment is usually restricted to finding the most comfortable positions and simple pain-relieving tablets that are safe in pregnancy. Gentle massage and the application of warmth may help, but manipulation and acupuncture should not be used. ●

67 I have had two children. My first pregnancy was fine but ever since my second childbirth, two years ago now, I have had a lot of low back pain. What is the cause and what can be done?

It is fairly common for women to have persistent back pain after childbirth. The cause is not always obvious. See your doctor to make sure there is no prolapse of the womb (see Q 64) or other abnormality causing your backache. If not, it is possible that something occurred either in the later months of pregnancy or during the delivery that has caused a mechanical problem in your back or has left active trigger points (see Q 58). It would be worth consulting a physiotherapist, osteopath, chiropractor or acupuncturist to see if such treatment would help. ●

My son, aged 15, has had a bad episode of backache. He is more or less over it now and he does not want to see the doctor, but I am worried. Should I insist?

As a general rule, the advice is that recent onset of back pain in people under 20 or over 50 needs to be investigated to make sure there is no serious cause for it. This is just a rule of thumb and it does not mean that people in these categories always, or even frequently, have some serious underlying disease. There does seem to be a tendency these days for back pain to occur in increasingly younger people, for unknown reasons.

That said, however, it would be just as well for your son to see the doctor. He may have some slight abnormality of the spine that makes his back unduly vulnerable (perhaps a congenital weakness of a vertebra, for example), or possibly Scheuermann's disease (see Q 69). It would be better to take measures to prevent recurrence at this stage rather than to wait until further episodes occur.

It is important to keep a sense of proportion about exercise for children and adolescents. There is a tendency for them to take less exercise nowadays than in the past, and there is concern that this may predispose them to health problems, including osteoporosis, in later life. They should certainly be encouraged to exercise, but their activities should also be monitored. For example, over-doing gymnastics in youth, especially in competition, can cause later spinal problems. During growth the bones and joints have not reached their full strength or stability and are vulnerable to damage. This does not, of course, mean that gymnastics are undesirable for a young girl or boy, but you should discuss the situation with the gymnastics teacher to make sure that the pupil is not being over-extended. If a young gymnast or other sportsperson does suffer an injury, you should seek the advice of a sports physiotherapist and make sure that he or she gets adequate rest and returns to full activity gradually and progressively. ●

69 **My son, aged 14, has been diagnosed as suffering from Scheuermann's disease. What is this caused by and is it serious?**

Scheuermann's disease (vertebral osteochondrosis) is due to prolapse of the soft disc material (nucleus pulposus) into the body of an adjacent vertebra, probably as a result of mechanical stress. When this happens the space between the vertebrae becomes narrowed and the vertebrae themselves become compressed and wedge-shaped. This leads to kyphosis (excessive spinal curvature). Usually, several vertebrae in the lower thoracic and upper lumbar regions are affected. The prolapsed disc material, called Schmorl's nodes, can easily be seen on CT scans.

The disease is more common in boys. It generally appears between the ages of 10 and 15. There is usually no pain. The abnormal vertebrae become bent outwards (kyphosis) and to compensate for this there is lordosis (concavity) of the lumbar spine below the level of the problem. This is, in fact, a form of hunchback. If left alone, the deformity may become static or may continue to get worse. Usually, progression can be prevented by means of corrective exercises prescribed by a physiotherapist, but some patients need in addition to wear a brace. In a few cases it is necessary to resort to surgery to produce spinal fusion, but fortunately this is rare.

Scoliosis refers to twisting of the spine. There are many different types of scoliosis. Some are due to other diseases affecting muscles or nerves and these are called secondary scoliosis. There is also idiopathic structural scoliosis, which means scoliosis that develops for unknown reasons.

There are three main types of structural scoliosis. One, which is more common in Europe than in North America, occurs in infants under the age of three and in boys more than girls. It usually affects the thoracic (chest) region and is convex to the left in most cases. Parents may notice the asymmetry, or they may see permanent skin folds on the concave side of the baby's chest. Fortunately, it clears up by itself in more than half the cases. If it does not, a corrective splint may be used; in rare cases surgery is carried out when the child is about 10 years old.

The second type, which is rare, affects older children, from four to nine years old. It seems to be a mixture of the infant type and the next group to be discussed, the adolescent.

Adolescent scoliosis principally affects girls. It takes the form of a lateral bend of the mid-thoracic spine, nearly always convex to the right and there is often a degree of spinal twisting as well. The deformity becomes more obvious if the child is asked to touch her toes. Unfortunately, this type of scoliosis does not clear up by itself and indeed it tends to become more pronounced with time. A physiotherapist will prescribe corrective exercises, perhaps together with a (Milwaukee) brace. X-rays are taken every three to six months to monitor progress. If the deformity is becoming worse it may be necessary to use serial plaster casts or traction. Some patients require surgery to insert corrective rods or carry out spinal fusion. ●

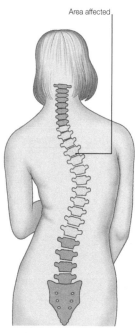

Area affected

Adolescent scoliosis affects the spine making it convex to the right.

SCOLIOSIS (TWISTED SPINE)

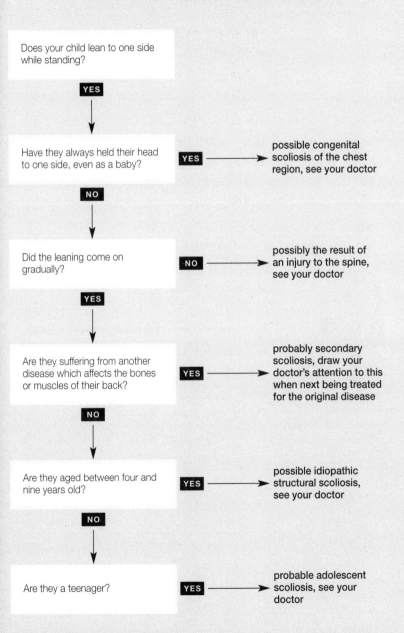

Does your child lean to one side while standing?

YES

Have they always held their head to one side, even as a baby? — **YES** → possible congenital scoliosis of the chest region, see your doctor

NO

Did the leaning come on gradually? — **NO** → possibly the result of an injury to the spine, see your doctor

YES

Are they suffering from another disease which affects the bones or muscles of their back? — **YES** → probably secondary scoliosis, draw your doctor's attention to this when next being treated for the original disease

NO

Are they aged between four and nine years old? — **YES** → possible idiopathic structural scoliosis, see your doctor

NO

Are they a teenager? — **YES** → probable adolescent scoliosis, see your doctor

Diagnosis & Treatment

The diagnosis of back pain has been radically transformed in recent years thanks to the development of new x-ray equipment. However, this doesn't mean that we now know all the causes; finding the cause, especially of chronic back pain, remains difficult. Surgery for back pain is sometimes, but by no means always, the right solution. This section considers the treatments on offer and what can be expected from them.

71 I had an x-ray of my back that showed a narrowing of the space between the fifth lumbar vertebra and the sacrum. What does this mean and should I worry?

Narrowing of disc space on x-ray indicates that there has been some loss of the jelly-like material (nucleus pulposus) from the disc in question, allowing the disc to become thinner. It does not necessarily produce any symptoms and it is impossible to tell how long ago this happened – it may have been many months or years previously.

The relation, if any, to symptoms is also quite uncertain. Because the distance between the vertebrae is reduced, it is possible that the nerve roots will become pinched at the point at which they emerge. No particular treatment is indicated for the intervertebral narrowing itself. If there are any back symptoms, they will be treated on their own merits.

I have had several attacks of back pain from which I haven't fully recovered. I had an x-ray that showed that everything was normal, but still the specialist wants me to have further tests. Why is this?

X-rays show only the bones – they do not show the soft tissues such as ligaments, discs and nerves. In particularly, they do not show whether there is a disc prolapse ('slipped disc', see Q 18) and if, so, whether it is pressing on the nerve roots or spinal cord. Special investigations are needed in order to reveal this.

In the past, the only way to diagnose soft tissue damage in the spine was to inject fluid into the spine that was opaque to x-rays. To do this, some local anaesthetic is injected into the back and then a special long needle is inserted between two vertebrae. This has to be done in the lower part of the lumbar spine to avoid damaging the spinal cord, which usually ends, in adults, at the lower border of the first lumbar vertebra. This is the same procedure as is used for a lumbar puncture. Once the fluid has been injected, an x-ray can be taken. Pressure on nerve roots, for example, from a prolapsed disc, will show up as an indentation on the opaque material. However, there is no way to extract this material and the use of the older oil-based substances sometimes led to problems such as arachnoiditis (see Q 52). Later, water-based materials were used instead to avoid this risk. Techniques of this kind have now largely been replaced by newer imaging methods, such as CT and MRI scans (see Q 73 and Q 74).

73 What is a CT scan?
Does it involve
x-rays?

A CT (computerized tomography) scan is a way of seeing the interior of the body in great detail. An ordinary x-ray shows everything as if it were in one plane, with the shadows of the various internal structures superimposed on one another. As a rule, x-rays are therefore taken in at least two planes, typically from the front or back and from the side and the radiologist then has to interpret the different appearances by combining them mentally. In an effort to improve resolution and to determine the depth of the different structures in the body, tomography was introduced: films were taken from a number of different directions and the results were combined to give something approaching a three-dimensional impression. ('Tomography' means 'slicing', the reference being to the different planes within the body that are visualized by the technique.)

A technician helps a patient as they enter the CT scanning apparatus.

With the introduction of computers this process became much more sophisticated and it is now possible to obtain excellent and very detailed pictures of internal structures, including the spine.

Some exposure to radiation is involved in CT scanning, but the dosage is small. Sometimes it is combined with myelography. ●

I'm going to be sent for an MRI scan of my back. What is it, how does it work and is it dangerous?

MRI stands for Magnetic Resonance Imaging. This is a relatively new technique that is still not available everywhere, partly because the apparatus is costly.

As the name implies, it works by means of a very strong magnetic field. You will not be able to feel this, but you will be asked to remove any objects, such as a watch, that could be affected by the magnetism. The magnetic field causes the nuclei of the atoms in your body to orient themselves in a particular direction in relation to the field. When the magnetic field is switched off the atoms lose this orientation and the rate at which they do so is measured. Because the various tissues in the body contain differing concentrations of the elements, it is possible to use the information to build up a picture of the structures in the body.

So far as is known at present, the procedure is entirely safe. However, it takes quite a long time to build up the picture and, because you have to remain totally stationary and in a small enclosed space, some patients do find that the procedure feels rather confined, even claustrophobic.

Both CT and MRI can produce astonishingly detailed pictures of the structures within the body. MRI, however, gives finer detail and it has the advantage that no X-rays are used, but MRI is less widely available and the cost of such imaging is still relatively high, although this is changing as more machines become available.

The use of medical imaging, as such examinations are called, has transformed the way in which back symptoms can be investigated and the cause of pain diagnosed. However, although the techniques are so powerful, the answers they give are not foolproof. The problem is that these increasingly sophisticated investigations are showing up more and smaller, abnormalities, but it is not always certain what these mean. In one recent study from the USA, for example, more than half of a group of people with no symptoms had some degree of abnormality on MRI scanning. The decision to operate, therefore, needs to be based on the results of physical examination and on the history, as well as on the findings of any scans that may be carried out.

75 When I was pregnant they used ultrasound to scan my womb and see what was happening to the baby. Why can't they use the same technique to see what is wrong with my back?

Ultrasound is another form of imaging in which high-frequency sound waves are used to generate a picture. Different tissues reflect the sound to different degrees. As far as is known, ultrasound is safe in pregnancy, which is why it is used to study the baby before birth. It is also used to study the pelvic organs in women at other times and it is used to study the heart and its valves. However, it generally does not work very well for the spine.

That said, ultrasound can be used to measure the diameter of the spinal canal in the lumbar and lower back regions, to see if it presses on the nerves in that area. This is relevant to the functioning of the legs and is used in the diagnosis of neurogenic claudication (*see Q 46*).

76 My right leg is weak following an episode of sciatica and the specialist wants me to have something called an EMG. What is this and why is it being done?

EMG stands for electromyography. This is a technique in which fine needles are inserted into the muscles and a recording is taken of the electrical activity within them. The reason your specialist is asking for this is to find out what your weakness is due to. EMG may be combined with studies of nerve conduction. Using these highly advanced methods the specialist can often identify the nerve root or roots that have been damaged. This approach works best when it is combined with the results of CT or MRI scans.

Although you will feel the insertion of the needles, the examination is neither particularly painful nor unpleasant to undergo.

77 I have been told
I am to have a
laminectomy. What
is this and how
does it work?

The laminae are the thin bony plates at the back of the vertebrae (see the illustration on page 8). Each vertebra has two laminae, one on each side; they meet in the middle to form an arch and the sequence of arches along the spine forms the spinal canal, which houses the spinal cord.

A laminectomy consists in cutting away all or part of a lamina or arch. The surgeon tries to do this as little as possible, in order to avoid exposing and weakening the spine unnecessarily. Sometimes it is possible to open the spinal canal through a window (fenestration), without cutting away any bone. Disc material that has leaked out from the annulus fibrosus is removed by an operation and bone that is pressing on nerves or nerve roots can be cut away. Sometimes a spinal fusion is carried out as well, to stabilize the spine. This can be done by inserting metal rods, screws or bone chips into the bone itself.

The surgeon's task is to open the spine sufficiently to see what is going on inside, but without leaving the spine in too unstable a state afterwards. Some surgeons have therefore tried a 'micro-surgical' technique, in which only a small opening is made and the whole operation takes well under an hour.

The success rate of surgery can be very good, especially for the treatment of pain, which is the main indication for the success of the operation. There is, however, generally less improvement in muscle strength. As with any operation there is, unfortunately, a failure rate and this poses a difficult dilemma for both patient and surgeon. Second and even third operations are sometimes carried out but the success rates diminish and the technical difficulties increase.

Recovery can also be difficult, with patients taking many weeks to return to normal.

78 A friend of mine has had discography performed to diagnose his back pain. Is this something I should ask my doctor to arrange for me?

Discography is a form of examination in which material opaque to x-rays is injected directly into a disc. This type of examination is particularly useful when diagnosing the presence of disc prolapse (*see Q 18*) and pressure on nerve roots. It is not widely used, however, because it does not give any more information than is provided by MRI scans. It is claimed that, if the injection reproduces the patient's pain, this proves that the disc in question is responsible for the symptoms, but there is uncertainty about this and about whether the information it yields is a good guide to the likely success of any subsequent operation.

79 I have had back pain for several months and now I have been advised to have an epidural injection. What is this and will it cure me?

This is a form of treatment in which a mixture of local anaesthetic and a corticosteroid drug is injected into the epidural space, between the tube formed by the dura mater (the outer lining of the spinal cord) and the spinal canal (the bony space within which the spinal cord is contained). This may be done either in the lumbar region, below the level of the spinal cord, or in the sacrum. The aim is to reduce inflammation and relieve pain. It does not provide a cure, but is intended to relieve pain while waiting for a natural recovery to occur. It is usually done as an outpatient procedure and patients can go home after 20 or 30 minutes. The pain may become increasingly worse for a day or two and then it improves. Some patients require a second treatment, or more, after one or two weeks.

Although epidural injections have been in use since the 1950s, there is little agreement about how they work or how useful they are. About a dozen studies have been carried out. The most recent, from Canada, concluded that although this treatment gave short-term relief, the eventual outcome was not affected and the need for surgery was not reduced.

30

My specialist wants to treat my back with facet joint injections. He didn't have much time to explain what this means. What are the facet joints and why should injecting them help?

This treatment is used to ease the pain in the lower back or neck thought to be caused by inflammation in the facet joints. Facet joints are the small weight-bearing joints at the back of the spinal column. Hydrocortisone is injected directly into the joint cavity of one or more facet joints, not unlike the procedure that is done in other osteoarthritic joints elsewhere in the body, such as the knee. The treatment can give long-lasting relief, although how it works is rather uncertain.

31

A specialist wants to do a sclerosing injection for my back because he says there is abnormal mobility. What does this involve and is it likely to help?

The rationale for treatment by sclerosing injection into the facet joints is that the pain is thought to be due to an instability of the invertebral joints in the back. An irritant (sclerosing) solution is injected into the ligaments around the spine with the aim of reducing the excessive mobility. Some patients find the treatment painful at first but it can give long-term relief. How it works is still uncertain but it is believed that the effect is actually similar to that of acupuncture (see Q 88).

82 I have been offered an injection of local anaesthetic in the back to help relieve my back pain. I would have thought that this would work for only a short time but the specialist says it may even give permanent relief. How could this work?

The idea of this treatment is to block the transmission of pain by means of the local anaesthetic. One would expect, as you say, that this would work for only a short time, until the anaesthetic wore off, but it can, in fact, give much longer relief. How this occurs is uncertain. Quite a few people think that the effect of local anaesthetic injections, like that of sclerosing injections (*see* Q 81), is produced by an acupuncture-like mechanism. In other words, what really matters is the fact that a needle has been inserted into the area, rather than any one property of the substance that is injected. ●

83 Can traction be used to successfully treat back or neck pain?

Traction is a method of treatment involving pulling on the spine in order to stretch it. This can be done manually by the therapist, or by using some form of harness attached to pulleys and weights. Often traction is applied for short periods, but sometimes patients are admitted to hospital for traction lasting several days, mainly for low back pain. However, the available evidence suggests that it has little value for the lower back. This is perhaps unsurprising, in view of the great strength of the lumbar spine and its ligaments, which would make it difficult to stretch this part of the spine.

Traction can be successful when applied to the neck and the thorax. Sometimes the relief is only short-lived, but other patients derive more lasting benefit. Some patients find that home traction is beneficial. A simple method for producing traction on the upper part of the body is to hang by the arms from a door or similar support, although the patient must have strong hands and a strong door is also required. More elaborate apparatus consists of frames to hold and tilt the body to provide traction by gravity, or harnesses that hold the head and apply traction via a pulley and cord with a weight attached. These methods should only be used on professional advice, since they are not always suitable and adopting a head-down position may be unsafe for some patients owing to the increase in blood pressure it causes in the head and eyes.

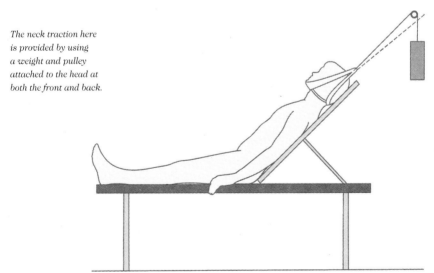

The neck traction here is provided by using a weight and pulley attached to the head at both the front and back.

I often see people wearing a surgical collar for neck pain and I know that some patients have corsets for back pain. Do these help?

Some patients certainly find collars and corsets helpful. They do not cure the disorder but they may give some relief from pain, perhaps by preventing movements of the neck.

Collars are generally made of latex sponge rubber or a similar material. Hard collars made of plastic are prescribed at times but are more uncomfortable. Whatever the material of which they are made, collars should be fitted individually to the patient. The main reasons for wearing a surgical collar are to provide support for the head and prevent movements that make the pain worse, to prevent jarring of the head while driving and to relieve pain that comes on when the patient is asleep.

Lumbar supports come in a bewildering variety of types. At one extreme there is the plaster-of-paris jacket, which totally encloses the body and prevents almost all movement and at the other end of the scale there is a simple affair consisting of little more than an elasticated belt; in between are various designs of corset reinforced with plastic or steel struts. A doctor would prescribe the style needed.

It is difficult to know what effect corsets have, if any. For some people, they have a strong placebo effect: they make them feel better for psychological reasons. Others, however, strongly dislike the idea of wearing a corset. Probably the main benefit of a

A soft surgical collar is commonly used for neck pain, especially in minor whiplash injuries.

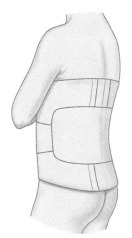

A lumbar support corset can be worn to reduce any strain on the lumbar region of the back.

corset is that it raises the pressure inside the abdomen. Research has shown, perhaps surprisingly, that this can decrease the load on the intervertebral discs and reduce the pressure in these by as much as 30 per cent. Another way of reducing pressure in the abdomen is by improving the strength and tone of the abdominal muscles and this is preferable.

Patients are often advised not to wear collars or corsets for too long in case it makes their muscles weak. However, there is no real evidence that this happens and research suggests that there is no weakening even after a corset has been worn for several years. Nevertheless, it is better not to become dependent on any kind of apparatus if this can be avoided.

A surgeon may prescribe a corset to be worn for a time after an operation to restrict movement, but this is generally for a fixed term only. ●

85 Lots of people these days keep talking about alternative medicine. Can this help back pain?

The term 'alternative' is rather misleading, because the boundary between conventional and unorthodox medicine is becoming increasingly blurred. Particularly in relation to back pain, physical therapies such as osteopathy, chiropractic and acupuncture are nowadays often recommended, or even used, by doctors. There is now a tendency to emphasize the way in which the different kinds of treatment can work together, so that formerly unorthodox therapies are frequently being described as 'complementary' rather than alternative.

One still frequent objection to unorthodox treatment is that acute backache generally gets better by itself even without treatment, so that it is unreasonable to claim that the improvement is due to the treatment. This is a fair point, but the same applies to much conventional treatment as well. Nowadays the tendency is to subject all kinds of treatment, both complementary and conventional, to objective scientific testing. This is done by means of controlled trials, in which the treatment to be tested is compared with other treatments or with dummy treatment and the outcome evaluated by statistical methods. There are practical difficulties in designing suitable trials for physical treatments, but some progress has been made and positive results reported for osteopathy, chiropractic and acupuncture.

Even if much complementary treatment works by means of the placebo effect, this is not necessarily a bad thing. Doctors have a tendency to think about it as being much the same as quackery, but there is a lot of evidence to show that the placebo effect is perfectly real. It may work in part by causing the release of naturally produced pain-relieving substances in the brain (the endorphins and enkephalins), or it may activate certain areas of the brain that can reduce the transmission of pain messages in the nervous system. The old idea that people who experience a strong placebo response are excessively suggestible is wrong – anyone may have this response in the right circumstance. The placebo response is generally supposed to depend on the strength of the patient's (and therapist's) belief in the treatment, but even this may be a half-truth. Some experiments seem to show that patients can experience pain relief even when they know they are receiving dummy treatment. In short, there is much that we still have to discover about the placebo effect.

From your point of view as a patient, it could certainly be worth trying one of the complementary approaches. However, it is important that you have realistic expectations about what can and cannot be achieved by this. Miracle cures are not on offer (and responsible therapists do not promise them), however, you may obtain relief of symptoms and, possibly, a quicker recovery than would be provided by nature unaided.

Acupuncture is often used as a means to treating back pain. The rounded needles ensure that they do not draw blood.

Are you suffering from back pain?

NO

YES

Is the pain worse when you are stressed?

YES → meditation
aromatherapy

NO

Is the pain the result of a repetitive movement, or of sitting or standing in the same position for long periods?

YES → psychotherapy
counselling
osteopathy

NO

Is the pain the result of stiffness in your neck, shoulders or arms?

YES → Alexander technique
chiropractic
osteopathy

NO

Do you have pain in your joints?

YES → Are you stiff in the morning and have increasing limitation in your movement?

YES → reflexology
hydrotherapy
t'ai chi chu'an

NO → osteopathy
chiropractic
shiatsu

NO

Do you have pain in your muscles?

YES → Are the muscles in your back tender to touch and does your back feel stiff?

YES → massage
hydrotherapy
yoga

NO

Does your back feel weak and do your muscles feel like they burn?

YES → Are your legs numb and do your muscles go into spasm?

YES → acupuncture
acupressure
osteopathy

NO → massage
shiatsu
acupuncture

There are indeed many kinds of complementary medicine, but we can classify them in a limited number of groups. The first group is made up of therapies involving some form of physical activity, often taking the form of hands-on treatment by the therapist. The principal examples are osteopathy, chiropractic and acupuncture. These forms of treatment are most suitable for so-called 'mechanical' back pain: that is, the kind that is made worse by movement and generally improves with rest. This means that these therapies could be worth considering if you suffer from osteoarthritis (see Q 60) or general 'wear and tear' in the spine. They can also help ankylosing spondylitis (see Q 39) as well as back pain of unknown cause where there is no question of a serious underlying illness.

The second group is made up of therapies that use medicines. The main forms are herbalism and homeopathy. Some patients with 'mechanical' pain may find these methods helpful, but on the whole they are more likely to help people whose back pain is part of a generalized illness, such as rheumatoid arthritis (see Q 60).

The third group of complementary treatments comprises the mental therapies, such as autogenic training and other forms of relaxation. These are generally most suitable for patients whose back pain is associated either with psychological tension or with depression.

There are other forms of alternative treatment, such as shiatsu and the Alexander technique, that are less easy to categorize; massage and aromatherapy also come into this category. Some of these are discussed in questions that follow.

Choosing the best treatment, therefore, depends mainly on what symptoms you are suffering from but also to some extent on your own preferences. For example, there is no point in considering acupuncture if you are terrified of needles. The placebo effect can occur with any form of treatment and should not be dismissed or ignored as worthless. All things being equal, you are more likely to derive benefit from a treatment you feel positive about. On the other hand, belief is not an essential prerequisite for success and there are many cases in which patients report success after treatments in which they initially had little confidence.

I'm thinking of looking for an complementary practitioner but I have no idea how to begin. Can you advise me?

In most cases the best place to begin is with your family doctor. Nowadays, doctors, especially family doctors, are often inclined to look favourably at some kinds of complementary therapy such as osteopathy, chiropractic, acupuncture and homeopathy, and some either practise these forms of treatment themselves or employ complementary therapists in their practice. Failing that, the doctor may know of a practitioner in the area who would be suitable.

If your doctor is unable or unwilling to help, you could ask for personal recommendations from friends or relatives, but be aware that what works for one person may be unsuitable or ineffective for another person.

Sometimes you may not know anyone who has been treated in this way and then you will have to rely on published lists and similar sources. In that case you should take certain common-sense precautions before agreeing to treatment. Do not be frightened or embarrassed to ask questions.

You should find out the cost of treatment and how long it will take. Do not agree to pay for long courses of treatment in advance – no responsible therapist will ask for this. No one should make a firm promise that you will be cured, because there is always an element of uncertainty at the outset. A responsible therapist should be willing to discuss the chances of success frankly with you, without making any over-optimistic claims. Remember that no form of treatment, conventional or complementary, has a 100 per cent success rate.

You should find out whether the therapist is a member of a professional body – such organizations exist for the majority of therapies. Ask about their training and how long they have been practising, but remember that time alone is not the only criterion. For example, a doctor practising the modern form of acupuncture may have taken only a short course of perhaps two weekends, compared with a traditional acupuncturist who has studied for some hundreds of hours, but the doctor starts from a basis of knowing anatomy and has been diagnosing and treating patients for many years, whereas the non-medical acupuncturist probably began training from scratch.

Be cautious if a therapist suggests that you stop taking prescribed drugs. You should only do this with the agreement of your doctor and the therapist should respect this proviso. In most cases, in fact, the therapist should write to your doctor, with your permission, to explain the treatment that he or she has in mind and what it is likely to achieve. Extreme hostility to conventional medicine on the part of the therapist is generally a bad sign.

I have heard good things about acupuncture for back pain but there seem to be different ways of doing it. Some practitioners put lots of needles in and leave them there for 20 minutes, while others put in very few needles for only a minute or two. I am confused; can you explain?

Acupuncture is a method of treatment that was practised in China for many centuries but has also been used in the West for quite a long time. The traditional Chinese version is based on an elaborate set of theories that form part of ancient Chinese philosophy. The main idea is that there are two opposed yet interwoven forces in the universe called yin and yang and that health depends on maintaining a balance between them. If there is too much yin or yang, the outcome is disease. There is also supposed to be a subtle kind of fluid or energy called chi (qi), which flows through the whole universe and also through the body, giving it life. The human organism is thought of as reflecting the organization of the universe. This is clearly a very different view of ourselves and of health and disease than that of modern science, but for some people this very difference is what makes it attractive.

Traditional acupuncture makes use of numerous acupuncture points (about 365 in all), most of which are said to lie on certain vessels or channels, usually called meridians in Western books. There are 59 of these meridians, but only 14 have acupuncture points on them. Great importance is attached to the exact location of the points, each of which is supposed to have particular effects when needled. Moreover, the points are not used in isolation but are combined in a complicated way based on the flow of chi through the meridians.

If you consult a traditional acupuncturist you will be asked questions about your way of life, emotional state and other matters. Next, the acupuncturist will inspect your tongue and take the pulses in both wrists, to help determine what kind of treatment to use. The needles will then be inserted in various acupuncture points; often about 20 needles will be used and they will be left in for about 20 minutes.

Moxibustion is a different but related form of treatment, in which certain herbs are burnt to produce heat. This may be applied directly to the skin, or the herbs may be wrapped round the needle to allow it to conduct the heat into the body.

Doctors and other health professionals, such as physiotherapists, who use acupuncture often think about it in terms of modern anatomy and physiology. If you are treated by them you will not receive a pulse or tongue diagnosis but will be assessed from a more conventional standpoint. The treatment may also be somewhat different; it will not be based on meridians or acupuncture points but is likely to make use of muscle trigger points (see Q 58). Fewer needles may be used and they may be left in for shorter periods, perhaps for just a minute or even less. Treatment with modern acupuncture is therefore often much quicker than the traditional version. Nevertheless it can work just as well and may even be more effective in some cases.

39 What kinds of problems can be treated with acupuncture? Is it safe? I'm worried about AIDS.

Acupuncture can be used to treat quite a wide range of disorders although it is by no means suitable for everything. Pain arising from the spine is one of the areas for which it is particularly applicable. It can be used to treat pain in the neck and lower back, as well as pain referred from those areas, for example to the shoulders, head or legs. In suitable cases about 75 per cent of patients will experience worthwhile relief; some need to continue with treatment more or less indefinitely, perhaps at intervals of eight to 12 weeks. As with all physical treatments, you should notice at least some improvement after two or three sessions; if nothing has happened by then, it may not be worth continuing.

If performed by a properly trained therapist, acupuncture is as safe as most other forms of treatment. So far, there are almost no reports of AIDS being transmitted by acupuncture. Hepatitis (infective jaundice) is a different matter; there is a real risk of this happening if contaminated needles are used. However, most acupuncturists now have disposable needles that are destroyed after use and provided this is done the risk of disease transmission is eliminated. The other danger is damage from the needles themselves. A properly trained therapist will have a good knowledge of anatomy; check that non-medical acupuncturists also have such knowledge.

90 I was thinking about trying acupuncture but I'm terrified of needles. I have heard about acupressure; would this do the same thing? And what about shiatsu?

Acupressure refers to the technique of pressing acupuncture points instead of needling them. It is supposed to work in the same way as acupuncture. Some doctors and physiotherapists think that the classic acupuncture points described in Chinese literature are in many cases the same as trigger points in Western medicine (see Q 58); that is, they are tender areas in muscles from which pain may radiate to other places in the body. Needling these sites can relieve pain, but simple pressure can do so too. This is acupressure.

Although acupressure can be effective, it has two disadvantages. One is that the pain relief often does not last very long. The other is that the pressure itself is generally painful; more painful, in fact, than needling. However, if you are really afraid of needles you may prefer the pain caused by pressure. Moreover, it is found that acupuncture generally does not work in people who are very afraid of it, so acupressure is worth trying in such cases.

Shiatsu, which literally means 'finger pressure', is a Japanese form of acupressure. The main difference is that it is carried out through loosely fitted clothes, with the patient lying on a thin mattress or on a massage table. The underlying theory is similar to that of traditional acupuncture and is based on concepts of 'energy flow'.

Many people these days seem to be having reflexology for their back pain and they say it works for them. What is it and how does it work?

Reflexology is a form of massage using the feet (and sometimes the hands). Like most other forms of physical therapy it is said to be ancient, with antecedents in China, but in its modern form it derives from the work early in the 20th century of William H. Fitzgerald, an American ear, nose and throat specialist. It is based on the idea that there are areas in the soles of the feet that relate to different parts and organs in the body. Reflexologists believe that by pressing various areas, 'energy blocks' are removed, allowing the body to heal itself.

The massage may be gentle, like a form of stroking, or firm, in which case it can be momentarily painful. You take off your shoes and socks, and possibly your trousers, and sit back on a treatment couch with your legs stretched out in front of you. The therapist may use talcum powder to allow the hands to move smoothly over your feet. An initial treatment usually lasts for an hour and a course of about six treatments is often recommended.

It has to be said that there is little or no scientific evidence to support the theory that there is a map of the body in the soles of the feet or that pressing them can relieve pain, but it is certainly true that many patients find the treatment pleasantly relaxing and beneficial. It is likely to work best for patients whose pain is associated with a lot of muscle tension.

92 Some of my friends say that massage helps their back pain. Is this a good treatment to try?

Massage, defined as the manipulation of soft tissues as a form of therapy, is probably one of the oldest forms of treatment. However, the modern techniques are largely based on methods developed in Sweden in the 19th century. People generally report that the experience is very pleasant and back pain is one of the principal reasons for which they have it. In some parts of the world it is customary for people to ask their relatives to walk on their back with bare feet in order to ease their back pain and simple forms of massage can certainly be carried out in the home. Professional masseurs use more elaborate techniques and a full body massage may last anything between 30 and 60 minutes. Although patients' reports of the effect of massage are often positive, there is little evidence to support its use. ●

93 I was thinking of seeing an osteopath but my mother says I should see a chiropractor instead. Now I'm confused. Which is more likely to help and how would their treatment differ?

Osteopathy and chiropractic had different origins but the methods they use and the disorders they treat are similar. Osteopathy derives from the work of an American doctor, Andrew Taylor Still (1828–1912); chiropractic from that of a Canadian, David Henry Palmer (1845–1913). (As a result of this difference, doctors today sometimes train as osteopaths but hardly any as chiropractors.)

The theories originally advanced to explain how the treatments worked were somewhat different, and so were the manipulative techniques. One continuing difference today is that chiropractors frequently make use of x-rays of the spine whereas osteopaths seldom do so. On the whole, however, the resemblances between them are more significant than their differences.

Contrary to what you may have heard, manipulation need not be painful or violent. Mostly it is gentle and pain-free. A wide range of techniques may be used: sometimes high-velocity thrusts, but often gentle massage-like movements to relax the muscles, or repetitive joint movements to improve the range of movement of a joint. Manipulative techniques are also used to some extent by physiotherapists.

At one time doctors did not accept the validity or effectiveness of these forms of treatment, but attitudes have now changed. The training of osteopaths and chiropractors is becoming regulated in a number of countries and the professions are becoming more closely allied to medicine.

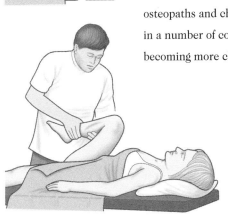

These images show how osteopathic treatments for back problems use gentle stretching of the muscles to relieve pain.

If you visit a therapist of this kind you will find that manipulation of the spine is only part of the treatment and not even the main part. Patients are generally expected to participate actively by modifying their lifestyle and activities. The main aim of treatment is to assess how well your body is functioning in relation to the demands made on it by your environment and personality. In the light of this assessment the therapist is able to offer advice about coping with the demands that you face. Manipulation has a part to play in this, but real improvement depends extensively on your cooperation.

Practitioners differ among themselves as regards the kinds of problems they think they can help, but acute and chronic back pain forms a major part of their work. However, there are some disorders in which manipulation could be actively harmful: for example, rheumatoid arthritis (*see Q 60*) affecting the spine or a spinal tumour. A properly trained therapist should recognize such disorders and refer them for treatment elsewhere, but even the best-trained people can make mistakes, which is one reason why it is generally advisable for you to consult your family doctor in the first instance. He or she will have all the relevant medical history and will usually be the best person to advise you on the suitability of physical treatment.

It would, of course, be unreasonable to expect a dramatic and complete cure from just one treatment, but there should normally be at least some improvement after two or three sessions. If no improvement at all has occurred by then, there is probably little point in continuing. As a rule, you should expect to receive between two and five treatments. After this, most patients are either pain-free or have reached a 'plateau' where further treatment produces no further improvement. At this point treatment should stop unless or until a relapse occurs. This pattern is very similar to that described for acupuncture.

94 **What about aromatherapy? How does this differ from ordinary massage?**

Aromatherapy combines massage with the use of various plant extract oils, which are aromatic and said to be therapeutic. The oils are supposed to be absorbed into the body both by inhalation and through the skin. A full body massage will last from one to two hours depending on your complaint and the practitioner. As in the case of ordinary massage without oils, people find the experience very pleasant and many find that it eases their back pain.

However, it is implausible to claim that anything that is rubbed on the skin can affect the underlying joints and muscles directly and there is virtually no objective evidence for the therapeutic effectiveness of the substances that are used. It is advisable to avoid this treatment in pregnancy, for fear of adverse effects on the unborn child from absorbing the aromatic substances.

Q

95 I'm pregnant and I'm getting a lot of backache. I know it is not a good idea to take drugs when you're pregnant, but I have heard that complementary medicine is safe because it is natural. Is this true?

The word 'natural' is rather meaningless in terms of treatment. There is nothing natural, for example, about sticking needles into people, as is done in acupuncture. In fact, physical methods of treatment, such as manipulation and acupuncture, should not as a rule be performed during pregnancy and certainly not in the first three months.

The same applies to the use of medicines, such as herbal preparations. Although it is true that herbs have been used for hundreds or even thousands of years, it does not follow that they are completely safe, especially in pregnancy. There may be a long time gap between taking a medication and the occurrence of any adverse effects and this can make it difficult to recognize the connection between the two. In general there is no such thing as a totally 'safe' medicine – anything that is capable of producing a good effect may also at times cause damage. The only totally safe medicine is one that is also totally ineffective.

In the case of herbal medicines, an extra complication comes from the fact that there is at present little or no control of their composition or purity. There have been many cases in which medicines, especially from the Far East, have been found to contain lead or other heavy metals. Some have contained conventional drugs such as corticosteroids (steroids), which explained their apparent effectiveness in relieving the pain of arthritis. Another danger is misidentification; there have been cases in which a toxic plant was used to make a herbal medicine by mistake, because its appearance or even its name was very similar to that of the medicinal plant.

Homeopathic medications are generally said to be safe in pregnancy because they are usually, although not always, used in a very dilute form. Nevertheless, this is no guarantee of safety. Most homeopathic practitioners believe that the medicines can cause temporary worsenings of patients' symptoms (aggravations) and if this is true it follows that there must be at least a potential risk to the unborn child. It is therefore preferable to avoid homeopathic medicines if possible, especially in the first three months of pregnancy.

When you are pregnant you should take exactly the same precautions in relation to alternative treatments that you would take in relation to conventional treatments: only use them if it is essential and always check with your doctor or specialist before agreeing to such treatment.

Q

96 I have been advised to try the Alexander technique to prevent recurrence of my back pain. Is it worth a go?

This is an approach to back problems that lies somewhere between the physical and the mental. It is concerned more with prevention than with the treatment of acute pain, but it could be useful for someone with long-term (chronic) back trouble. Its founder, F. Matthias Alexander, was an Australian actor who began to suffer from difficulties with his voice. He concluded that this was because he was inadvertently adopting an abnormal head posture during his performances and this led him to study the way in which he used his spine. Eventually he formulated the system that bears his name.

Teachers of the Alexander technique work in a variety of ways, but the general idea is that most of us have unconsciously adopted inefficient or positively harmful ways of using our bodies, especially our spines. The Alexander teacher aims to restore what ought to be the natural way of sitting, standing and moving. Initially this has to be done consciously, by continually giving oneself reminders of what one ought to be doing, but eventually the body adapts to the 'new' patterns of use and they become automatic. The teaching is generally done individually and regular practice is needed to change a lifetime's habits. Lessons last from 30 to 45 minutes and most people require between 15 and 30 lessons to build up the new patterns of movement.

I suffer a lot from stress at work and someone suggested that this is a cause of the pain in my neck and back. They also advised me to try autogenic training, which they found very helpful. What is autogenic training anyway?

It is true that stress does give rise to pain in the neck and back in many people. This is usually ascribed to tension in the neck and back muscles, which is undoubtedly part of the explanation. However, there is more to it than this. The work environment can be unsatisfactory in various ways. These include simple mechanical matters, such as the position you sit in at your desk, the height of your chair or the angle of your computer screen, but they also include psychological pressures of different kinds. Perhaps you get on badly with your boss, or perhaps you do not care for the work you have to do; perhaps you are a teacher, with the prospect of an inspector's visit weighing on your mind. It is generally agreed that for many people work is becoming increasingly stressful these days, for a variety of reasons.

It may not be possible for you to alter many of these circumstances, in which case the best solution is to try to improve your own inner resources for coping with stress. Autogenic training does often help in this way. It was developed early in the 20th century by a German psychologist and neurologist called Johannes Schultz. It is a detailed programme of relaxation, which is similar in some ways to self-hypnosis and also has affinities with yoga. It is usually taught in groups. You learn a series of mental exercises that aim to turn off the body's response to stress. These begin by directing the attention to various parts of the body: arms, legs, head, chest, abdomen and so on. Each session lasts at least an hour, sometimes more, and a basic course consists of eight sessions. In between sessions you practise the techniques at home and keep a diary of what you feel while doing them. You are also taught simple exercises that you can do during the day if you feel yourself coming under stress either at home or at work. All this is quite demanding in terms of both time and concentration and you should not embark on the programme unless you are prepared to work at it seriously.

Three months after the initial course you are asked to return for review to make sure that things are progressing satisfactorily. There are also more advanced methods of autogenic training that you can explore if you wish to do so.

Teachers, who are mostly health professionals of one kind or another, undergo a rigorous training and will carry out an assessment of your state of health before starting you on the programme. ●

Q

98 Autogenic training
isn't available locally.
Can I learn it from a
book or is there
something similar I
could do for myself?

It is not possible to learn autogenic training from
a book but a number of other possibilities do
exist. There are many books on mental relaxation
and you may find one of these to be helpful. There
are also tapes of relaxation instructions, which are
really a form of self-hypnosis. If you prefer to attend
formal classes, these are often available at a local
hospital and frequently form part of the work of
ocupational therapists (see Q 51). Still another idea
is to take up some form of meditation; there are
many varieties of this.

One simple meditation technique that anyone can
try for themselves is just to set a few minutes aside
each day to sit quietly, perhaps with closed eyes, and
to allow the attention to rest on any feelings of strain
or tension there may be in the body, five or ten
minutes of this are enough at the beginning. The
idea is not to try to get rid of the tensions by effort,
or to do anything with them, but simply to
experience them and see what they feel like. One
could think of the tension as a message that the body
is giving – a form of complaint – and the aim of the
meditation is simply to become aware of this,
without trying to push it aside.

If you do this, you will probably find that there is
a great deal of tension in the muscles of your face,
neck and shoulders particularly. As soon as this

comes into your awareness, you can 'let go' of it; your face relaxes and your shoulders droop a little. Before long you find your attention has wandered onto something else – perhaps some worry that you cannot get out of your mind. Now check your muscles again. Perhaps they have tensed up once more. In that case, just let the tension go as you did before. Or perhaps you are now thinking of some quite neutral matter and your muscles are still relaxed; in that case you can stop meditating and return to your normal activities.

As you continue with this simple meditation on a daily basis, you will probably find that you become aware of the tensing up of your muscles even outside the formal meditation periods; then you can let go of it immediately and in this way the cycle of tension and pain no longer builds up so intensely.

99 My friend says homeopathy has helped his back. Is it worth trying?

This form of treatment originated in the early 19th century, due to the research of Samuel Christian Hahnemann (1755–1843). He was a conventional doctor who became understandably disillusioned with the medical practices of his day and cast about for something better. After many years of trial and error he arrived at the system he later called homeopathy, meaning 'treating like with like'. It has something in common with herbalism; most of the medicines available in Hahnemann's time were herbal, but the principle of selection is different. The central idea of homeopathy is to choose a medicine that, when given to a healthy person, produces symptoms as similar as possible to those of the disease to be treated.

In the case of back pain, for example, a commonly used medicine is *Rhus toxicodendron*, or poison ivy. This is a plant that causes severe itching and blistering of the skin in people who are sensitive to it. When taken by mouth it also causes aching in the back and joints, so it is said to be 'homeopathic' to both skin rashes, and back and joint pain.

Another of Hahnemann's innovations was the use of very small doses of medicines. At first he did this to avoid unwanted toxic effects, but later he came to believe that these small doses, when prepared in a special way involving repeated grinding or hard shaking, actually made the medicines more active than they would otherwise be. The great majority of homeopathic medicines are still given in a highly dilute form and are prepared using the process that Hahnemann called dynamization or potentization.

The aim of a homeopathic consultation is to select the right medicine for the patient as an individual, rather than for the disease. A homeopathic prescriber takes a detailed history, paying attention to the way the symptoms behave, how they progress, the time of day they are worst and what makes them better or worse. Importance is also attached to how the patient feels emotionally about the illness. Many other factors may be taken into account, including weather and food preferences. In the case of *Rhus toxicodendron*, the features that might suggest its use would be an improvement in the symptoms in response to warmth and initial stiffness and pain that wear off after a short period of activity; flitting pains here and there, together with tearfulness, might suggest a different medicine. The medicines are given by mouth, usually as small sugar pills or powders. Sometimes just a few doses are given and the patient is then watched for some weeks to see what effect follows, or else the medicine may be given two or three times daily for several weeks. In mainstream homeopathy, medicines are given singly, but there are several different ways of practising homeopathy and in Germany it is common to give mixtures of homeopathic medicines.

Homeopathy is probably unlikely to help back pain that is due to mechanical problems, but it is worth trying for pain due to rheumatoid arthritis or similar generalized diseases.

**Is herbalism the
same as homeopathy
and can it help
back pain?**

Herbalism is not the same thing as homeopathy,
although there is some overlap; certain
medicines are used by both forms of treatment.
Unlike most homeopathic medicines, however,
herbal preparations are not given in a dilute form.

Almost all the medicines used by doctors are
herbal in origin. Some are still in use today, but the
main difference from modern medicine is that
herbalists use the original plant, whereas modern
drugs contain a single purified isolated chemical.
Herbalists claim that plants used in their natural
state work with fewer side-effects. They believe that
herbs correctly prescribed can restore the disturbed
balance of the body. Diet is also used for this purpose.

It is commonly believed that herbal medicines are
entirely safe because they are 'natural'. This is not
always true, herbalism uses significant doses of the
substances, rather than the highly dilute preparations
used in homeopathy, so there is a potential risk of
poisoning if you prescribe herbal medicines for
yourself. Mis-identification of plants has led to some
episodes of poisoning and some imported herbal
medicines have been found to contain heavy metals
or steroids. Buying herbal medicines over the
counter must therefore be done carefully and if you
consult a herbalist for treatment you should make
inquiries to ensure that they are adequately trained.

Useful Information

FURTHER READING

LIOC BUM, *Back and Neck Pain: The Facts,* Oxford University Press, 1999

BROWNSTEIN, ART, *Healing Back Pain Naturally – The Mind-Body Programme Proven to Work,* Newleaf/Harbour Press, USA, 2000

FISHMAN, SCOTT, *The War on Pain – Turning the Tide Against Suffering,* Newleaf/Harper Collins, USA, 2000

MELZACK, RONALD and WALL, PATRICK, *The Challenge of Pain* (Revised Edition), Penguin Books, 1996

WALL, PATRICK, *Pain: The Science of Suffering,* Weidenfeld and Nicolson, 2000

ANON, *Understanding Back Trouble,* A Consumer Publication. Which? Books, 1991

USEFUL ADDRESSES

Arthritis Care
18 Stephenson Way, London NW1 2HD
tel: 020 7916

The Arthritis Foundation of Ireland
1 Clan William Square
Grand Canal Quay, Dublin 2, Ireland
tel: 01 661 8188
website:
http://www.arthritis-foundation.com

British Acupuncture Council
Park House
206–208 Latimer Road
London W10 6RE
tel: 020 87350400
e-mail: info@acupuncture.org.uk

British Chiropractic Association
Blagrave House, Blagrave Street,
Reading RG21 1QB
tel: 0118 950 5950

British School of Osteopathy
275 Borough High Street
London SE1 1JE
tel: 020 7407 0222

The Chartered Society
of Physiotherapy
14 Bedford Row
London WC1R 4ED
tel: 020 7306 6663/4/5

The Society of Teachers
of the Alexander Technique
20 London House
266 Fulham Road
London SW10 9EL
tel: 020 7351 0828
e-mail: enquiries@stat.org.uk

Chiropractor's Association
of Australia (National) Ltd
Suite 4, 148 Station Street
Penrith NSW 2750
tel: 02 4731 8011
website:
http://www.caa.com.au

WEBSITES

BackCare, the National Organization
for Healthy Backs
http://www.backpain.org

BMAS (acupuncture information):
http://www.medical-acupuncture.co.uk

Back pain in children
http://www.vh.org/Providers/Textbooks
/BackPainInChildren/BackPainChildren.
html

Low back pain and sciatica
http://www.voiceoftheinjured.com/
a-si-back-spine-pain-sciatica-diagnosis-
treatment.html
http://www.sechrest.com/mmg/back/
backpain.html

National Center for Complementary
and Alternative Medicine
http://www.nccam.nih.gov/nccam/fcp/
factsheets/acupuncture/acupuncture.htm

Neck pain
http://www.wadhurst-physio.co.uk/
neck%20pain.htm
http://www.spineuniverse.com/treatment/
nonsurgical/physiatrist_neck.html

Pain at the back of the head
http://www.netdoctor.co.uk/diseases/
facts/neckpains.htm

Physiotherapy and occupational therapy
www.shef.ac.uk/~nhcon/phys.htm

Scoliosis, osteoporosis, Paget's disease
http@//yourhealth.queens.org/Health
Topics/

Index